Using Children's Literature to Teach Problem Solving in Math

Learn how children's literature can help K–5 students see the real-life applications of mathematical concepts. This user-friendly book shows how to use stories to engage students in building critical reasoning, abstract thinking, and communication skills, all while helping students understand the relevance of math in their everyday lives. Each chapter is dedicated to one of the eight Standards for Mathematical Practice, and offers examples of children's literature that can be used to help students develop that practice.

You'll find out how to:

◆ Encourage students to persevere in solving mathematical problems and use multiple approaches to find the answer;
◆ Help students reason abstractly with the aid of concrete objects and visuals;
◆ Guide students in constructing arguments to explain their reasoning and engage in critical discussion with their peers;
◆ Teach students to recognize mathematical patterns and use them to solve problems efficiently;
◆ And more!

The book offers activities for beginners as well as for more advanced problem solvers. Each chapter also provides guidance for ELLs and students with special needs, so no matter your classroom environment, you'll be able to use these strategies to make math class more dynamic, engaging, and fun.

Jeanne White has been an educator since 1992 when she began teaching elementary school in the south suburbs of Chicago. She is currently a Professor of Education at Elmhurst College.

Other Eye On Education Books, available from Routledge

(www.routledge.com/eyeoneducation)

Guided Math in Action:
Building Each Student's Mathematical Proficiency with Small-Group Instruction
Nicki Newton

Math Workshop in Action:
Strategies for Grades K–5
Nicki Newton

Math Running Records in Action:
A Framework for Assessing Basic Fact Fluency in Grades K–5
Nicki Newton

Creating a Language-Rich Math Class:
Strategies and Activities for Building Conceptual Understanding
Sandra L. Atkins

Performance Tasks and Rubrics for Early Elementary Mathematics, Second Edition:
Meeting Rigorous Standards and Assessments
Charlotte Danielson and Pia Hansen

Performance Tasks and Rubrics for Upper Elementary Mathematics, Second Edition:
Meeting Rigorous Standards and Assessments
Charlotte Danielson and Joshua Dragoon

The Mathematics Coaching Handbook, Second Edition:
Working with K–8 Teachers to Improve Instruction
Pia M. Hansen

Math Intervention P–2, Second Edition:
Building Number Power with Formative Assessments, Differentiation, and Games, Grades PreK–2
Jennifer Taylor-Cox

Math Intervention, 3–5, Second Edition:
Building Number Power with Formative Assessments, Differentiation, and Games, Grades 3–5
Jennifer Taylor-Cox

Using Children's Literature to Teach Problem Solving in Math

Addressing the Standards for Mathematical Practice in K–5

Second Edition

Jeanne White

Routledge
Taylor & Francis Group

NEW YORK AND LONDON

Second edition published 2017
by Routledge
711 Third Avenue, New York, NY 10017

and by Routledge
2 Park Square, Milton Park, Abingdon, Oxon OX14 4RN

Routledge is an imprint of the Taylor & Francis Group, an informa business

© 2017 Taylor & Francis

The right of Jeanne White to be identified as author of this work has been asserted by her in accordance with sections 77 and 78 of the Copyright, Designs and Patents Act 1988.

All rights reserved. No part of this book may be reprinted or reproduced or utilised in any form or by any electronic, mechanical, or other means, now known or hereafter invented, including photocopying and recording, or in any information storage or retrieval system, without permission in writing from the publishers.

Trademark notice: Product or corporate names may be trademarks or registered trademarks, and are used only for identification and explanation without intent to infringe.

First edition published by Routledge 2013

Library of Congress Cataloging in Publication Data
Names: White, Jeanne (Jeanne Marie)
Title: Using children's literature to teach problem solving in math :
    addressing the standards for mathematical practice in K-5 / by
    Jeanne White.
Description: New York : Routledge, 2017. | Includes bibliographical
    references.
Identifiers: LCCN 2016008696| ISBN 9781138694705 (hardback) |
    ISBN 9781138694712 (pbk.) | ISBN 9781315527536 (e-book)
Subjects: LCSH: Mathematics–Study and teaching (Elementary) |
    Mathematics–Study and teaching (Elementary)–Standards. |
    Language arts (Elementary) | Language arts–Correlation with
    content subjects.
Classification: LCC QA135.6 .W479 2017 | DDC 372.7/049–dc23
LC record available at http://lccn.loc.gov/2016008696

ISBN: 978-1-138-69470-5 (hbk)
ISBN: 978-1-138-69471-2 (pbk)
ISBN: 978-1-315-52753-6 (ebk)

Typeset in Palatino
by Wearset Ltd, Boldon, Tyne and Wear

Printed and bound in the United States of America by Publishers Graphics, LLC on sustainably sourced paper.

Contents

Acknowledgments x
About the Author xi

Introduction: Addressing the Standards for Mathematical Practice in K–5 1
Breaking Down the Practice Standards 2
Creating a Problem Solving Community 4

1 Make Sense and Persevere: SMP 1—Make Sense of Problems and Persevere in Solving Them 6
What Does This Standard Mean for Grades K–2 Problem Solvers? 6
Seven Blind Mice by Ed Young
K.OA Decompose numbers into pairs 7
MATH-terpieces: The Art of Problem Solving by Greg Tang
1.OA Use two and three addends to find a sum 11
Splash! by Ann Jonas
2.OA Represent addition and subtraction problems 14
What Does This Standard Mean for Grades 3–5 Problem Solvers? 16
The Warlord's Kites by Virginia Walton Pilegard
3.MD Use the formula for the area of a rectangle 16
A Remainder of One by Elinor J. Pinczes
4.OA Solve multi-step word problems with whole numbers 18
Multiplying Menace by Pam Calvert
5.NF Multiply a whole number by a fraction 22
Wrapping It Up 26

2 Reason Abstractly: SMP 2—Reason Abstractly and Quantitatively 27
What Does This Standard Mean for Grades K–2 Problem Solvers? 27
Each Orange Had 8 Slices by Paul Giganti, Jr.
K.OA Count to answer, How many? 28

Ten Flashing Fireflies by Philomen Sturges
1.OA Explore the Commutative Property of Addition 30
Rooster's Off to See the World by Eric Carle
1.OA Explore the Associative Property of Addition 32
Spaghetti and Meatballs for All! by Marilyn Burns
2.G Partition rectangles into rows and columns. 34
What Does This Standard Mean for Grades 3–5 Problem
Solvers? 37
Racing Around by Stuart J. Murphy
3.MD Solve problems involving perimeters of polygons 37
Fractions, Decimals and Percents by David A. Adler
4.NF Understand decimal notation for fractions 39
Count to a Million by Jerry Pallotta
5.NBT Understand the value of 0 in a multi-digit number 41
Wrapping It Up 43

**3 Construct Arguments: SMP 3—Construct Viable Arguments
and Critique the Reasoning of Others** 44
What Does This Standard Mean for Grades K–2 Problem
Solvers? 44
How Many Mice? by Michael Garland
K.OA Represent addition and subtraction 45
Mall Mania by Stuart J. Murphy
1.NBT Add two-digit numbers using various strategies. 47
How Many Seeds in a Pumpkin? by Margaret McNamara
2.NBT Skip-count and compare three-digit numbers. 48
What Does This Standard Mean for Grades 3–5 Problem Solvers? . . . 52
Full House by D.A. Dodds
3.NF Understand how fractions are written as part/whole 53
The Warlord's Puzzle by Virginia Walton Pilegard
4.G Classify 2-D figures and recognizing right triangles. 55
The Hershey's Milk Chocolate Fractions Book by Jerry Pallotta
5.NF Use equivalent fractions to add and subtract fractions 58
Wrapping It Up. 60

4 Create a Model: SMP 4—Model with Mathematics 61
What Does This Standard Mean for Grades K–2 Problem
Solvers? 61
The Doorbell Rang by Pat Hutchins
K.OA Represent addition in various ways 62

Bigger, Better, Best! by Stuart J. Murphy
1.G Compose 2-D shapes into composite shapes. 66
Alexander, Who Used to be Rich Last Sunday by Judith Viorst
2.MD Solve word problems with money. 71
What Does This Standard Mean for Grades 3–5 Problem Solvers? . . . 74
Tiger Math by Ann Whitehead Nagda and Cindy Bickel
3.MD Draw graphs to represent a data set with several categories. . . . 75
Actual Size by Steve Jenkins
4.MD Make a line plot to display a data set of measurements. 78
Polly's Pen Pal by Stuart J. Murphy
5.MD Convert among different-sized measurement units. 81
Wrapping It Up. 83

5   **Use Mathematical Tools: SMP 5—Use Appropriate Tools Strategically** 84
What Does This Standard Mean for Grades K–2 Problem Solvers? 84
Patterns in Peru by Cindy Neuschwander
K.G Describe relative positions 85
How Big is a Foot? by Rolf Myller
1.MD Iterate length units. 88
Sir Cumference and the Sword in the Cone by Cindy Neuschwander
2.G Recognize attributes of 3D objects 91
What Does This Standard Mean for Grades 3–5 Problem Solvers? 94
Inchworm and a Half by Elinor Pinczes
3.MD Generate measurement data by measuring lengths 95
Sir Cumference and the Great Knight of Angleland by Cindy Neuschwander
4.MD Understand concepts of angles. 97
Perimeter, Area and Volume by David A. Adler
5.MD Understand concepts of volume measurement 101
Wrapping It Up. 104

6   **Attend to Precision: SMP 6—Attend to Precision**. 105
What Does This Standard Mean for Grades K–2 Problem Solvers? 105
If You Were a Triangle by Marcie Aboff
K.G Identify 2D shapes. 106
Lemonade for Sale by Stuart J. Murphy
1.MD Represent and interpret data. 108

Measuring Penny by Loreen Leedy
2.MD Measure and estimate lengths. 112
What Does This Standard Mean for Grades 3–5 Problem
Solvers? 115
Chimp Math by Ann Whitehead Nagda and Cindy Bickel
3.MD Solve problems involving measurement of intervals
of time 115
Hershey's Milk Chocolate Weights and Measures by Jerry Pallotta
4.MD Express and record units of measurement 119
Sir Cumference and the Viking's Map by Cindy Neuschwander
5.G Graph points on a coordinate plane 122
Wrapping It Up. 124

7   **Look for Structure: SMP 7—Look for and Make Use of
Structure** 125
What Does This Standard Mean for Grades K–2 Problem
Solvers? 125
The Button Box by Margarette S. Reid
K.MD Identify attributes for sorting. 126
The Greedy Triangle by Marilyn Burns
1.G Explore attributes of shapes 129
Earth Day— Hooray! by Stuart J. Murphy
2.NBT Use place value to add and subtract. 133
What Does This Standard Mean for Grades 3–5 Problem
Solvers? 137
Hershey's Kisses from Addition to Multiplication by Jerry Pallotta
3.OA Relate multiplication and division 138
If You Hopped Like a Frog by David M. Schwartz
4.MD Solve problems involving measurement. 141
Cheetah Math by Ann Whitehead Nagda
5.NBT Explain division calculations 144
Wrapping It Up. 147

8   **Apply Repeated Reasoning: SMP 8—Look for and Express
Regularity in Repeated Reasoning** 148
What Does This Standard Mean for Grades K–2 Problem
Solvers? 148
Bunches of Buttons: Counting by Tens by Michael Dahl
K.CC Count to 100 by tens 149
The King's Commissioners by Aileen Freidman
1.NBT Represent tens and ones. 152

Sir Cumference and All the King's Tens by Cindy Neuschwander
2.NBT Represent three-digit numbers. 154
What Does This Standard Mean for Grades 3–5 Problem
    Solvers? 157
Math Appeal by Greg Tang
3.OA Represent and solve problems involving multiplication. 158
The Warlord's Beads by Virginia Walton Pilegard
4.NBT Understand place value for multi-digit numbers. 161
Anno's Magic Seeds by Mitsumasa Anno
5.OA Generate numerical patterns 163
Wrapping It Up. 166

Next Steps 167
Appendix 168
References 170

Acknowledgments

I would like to thank my family, who have consistently supported everything I do both personally and professionally. I would like to thank my high school math teachers at Thornwood High School, who were all excellent teachers and were the reason for my love of mathematics. I have to thank my editor, Lauren Davis, who discussed ideas for this book aimed at Kindergarten through fifth grade teachers.

About the Author

Dr. Jeanne White has been an educator since 1992 when she began teaching elementary school in the south suburbs of Chicago. She earned her doctorate in Curriculum & Instruction in 2003 and began her career at Elmhurst College as part of the full-time faculty in 2005. She teaches the math methods courses for the teacher candidates in the early childhood and elementary education programs and works with in-service teachers in the Master of Education in Teacher Leadership graduate program. She is currently the chairperson of the Department of Education.

She has presented internationally, nationally and locally on topics of math education, specializing in elementary school mathematics. She has conducted workshops for teachers in Australia and South Africa on how to use everyday objects to facilitate early mathematics instruction. She does consulting for school districts on implementing the Common Core State Standards for Mathematics and has written numerous articles on math education.

Introduction
Addressing the Standards for Mathematical Practice in K–5

At the root of the Common Core State Standards for Mathematics are the eight Standards for Mathematical Practice (SMP), describing "varieties of expertise that mathematics educators at all levels should seek to develop in their children" (CCSSI 2010, p. 6). As students learn skills and strategies for performing calculations, they must also acquire skills and strategies for proving how they can reason, communicate, represent and make connections as they solve mathematical problems. Unlike many of the previous state standards used by school districts around the nation, the Common Core State Standards for Mathematics include specific standards for students at the Kindergarten level. This provides opportunities for teachers to instill problem solving skills at an early age which can set the foundation for critical thinking in mathematics throughout the elementary grades.

Students in Kindergarten, first and second grade may have difficulty understanding the language of the eight SMP as they are written in the Common Core State Standards for Mathematics. Teachers can get together with colleagues within their school or district to rewrite each SMP in student-friendly language. One example of how the SMP can be rewritten for students in Kindergarten through second grade is shown in Table I.1.

Introducing students to mathematical problems can be challenging considering students in Kindergarten and first grade are at early stages of learning to read, write, add and subtract. One way teachers can facilitate the teaching of word problems is to use children's literature as the context. In this way, teachers can show illustrations from the book, use characters from the story and discuss any unfamiliar vocabulary before or during the presentation of the book. When teachers incorporate children's literature into mathematics, opportunities arise for young children to see math in their own lives. If they see how characters in a story use math to solve problems, they can better understand how people around them use math to solve everyday problems. The use of literature can also facilitate problem solving by presenting a common language and structure for the teacher and students to use while they engage in the math content.

Each chapter in this book is dedicated to one of the eight SMP with examples of how activities based on children's literature can be used as a way to apply essential aspects of each standard. The activities are all based on the

Table I.1   K-2

SMP	What This Means
1. Make sense of problems and persevere in solving them	*I can explain the meaning of the problem in my own words.*
2. Reason abstractly and quantitatively	*I can understand how the numbers in the problem are related.*
3. Construct viable arguments and critique the reasoning of others	*I can listen and respond to the way others solved the same problem.*
4. Model with mathematics	*I can identify important quantities and represent their relationships.*
5. Use appropriate tools strategically	*I can learn how to use different mathematical tools.*
6. Attend to precision	*I can explain how I solved a problem using mathematical terms.*
7. Look for and make use of structure	*I can find a pattern in a problem.*
8. Look for and express regularity in repeated reasoning	*I can look for repeated calculations.*

Standards for Mathematical Content in Kindergarten through fifth grade. Teachers do not have to start with Chapter 1 but should use activities from any chapter that would be relevant for their current students. Kindergarten, first and second grade teachers might focus on one SMP a month or may just choose a few on which to focus for the year. Third through fifth grade teachers could build on their students' knowledge of the SMP from previous grades, focusing on one or two each month, while helping their students connect the ideas reflected in each SMP so they are not viewed in isolation.

Breaking Down the Practice Standards

Students may have difficulty relating to the language of the eight SMP as they were originally written in the Common Core State Standards for Mathematics. Students may also feel overwhelmed trying to master all of the skills described in each practice standard. Teachers may find it useful to rewrite each SMP in student-friendly language for students in intermediate grades as well, or to solicit the advice of students for ways to break down these standards into manageable tasks. One example of how each SMP can be broken down into multiple tasks for students in third through fifth grade is shown in Table I.2.

I can statements...

Table I.2   3-5

SMP	What This Means – children's terms
1. Make sense of problems and persevere in solving them	◆ *I can restate the meaning of the problem and create a plan to reach a solution.* ◆ *I can analyze what is given, what is not given and the goal of the problem.* ◆ *I can use verbal descriptions, tables, graphs and diagrams to solve the problem.* ◆ *I can determine whether or not the answer makes sense by using a different method.*
2. Reason abstractly and quantitatively	◆ *I can understand how the quantities in the problem are related.* ◆ *I can represent the context of the problem in order to create an equation.* ◆ *I can apply the properties of operations while solving the problem.*
3. Construct viable arguments and critique the reasoning of others	◆ *I can use assumptions, definitions and prior results to construct my argument.* ◆ *I can analyze the arguments of others in response to the way they solved the same problem.* ◆ *I can create counterexamples and identify flaws in an argument.*
4. Model with mathematics	◆ *I can make connections between the problem and its representation in everyday life.* ◆ *I can identify important quantities and represent their relationships.* ◆ *I can use inferences and estimates to simplify the problem.*
5. Use appropriate tools strategically	◆ *I can determine when it is appropriate to use specific mathematical tools.* ◆ *I can identify relevant mathematical resources to pose or solve the problem.* ◆ *I can recognize strengths and limitations of various tools.*
6. Attend to precision	◆ *I can apply definitions and state the meaning of symbols when I communicate my answer.* ◆ *I can correctly apply labels and specify units in my answer.* ◆ *I can use precise language to explain how I solved the problem.*
7. Look for and make use of structure	◆ *I can find a pattern or structure within the problem.* ◆ *I can determine the significance of information in the problem.* ◆ *I can step back and look at the problem in a new way, making adjustments as needed.*
8. Look for and express regularity in repeated reasoning	◆ *I can look for repeated calculations, general methods or shortcuts while solving the problem.* ◆ *I can recognize when a general formula exists to help solve the problem.* ◆ *I can make conclusions about my results as I work through a possible solution.*

Creating a Problem Solving Community

Set the stage for problem solving in your classroom by reading excerpts from the book *The Math Curse* (1995) by Jon Sciezka. In this story, the narrator is plagued by looking at every part of life as a math problem—eating cereal, getting dressed and moving through the school day. Challenge your students to think of their day as a series of math problems. What time do they have to leave the house in order to get to school on time? How many boys and girls are in the class? How many rode the bus? How many did not? Start a bulletin board for questions created by students and choose one problem to solve each day. Share events from your life that can be put into a word problem. Communicate to your students that everyone is a problem solver, including yourself.

Decorate your classroom in a way that demonstrates the importance of mathematics to students and others who walk into your classroom. Post a Math Word Wall with each of the mathematical terms used at your grade level, with a definition, picture or example. There should be posters on the wall for mathematical concepts such as fractions, place value and various types of graphs. In the classroom library, there should be children's literature that includes math concepts, such as the titles used in this book. Math instruction should not be restricted to a specific time of day but should be practiced throughout the day. Students can organize their class schedule and assignments in a chart, estimate how long a task may take to complete, help the physical education teacher calculate the number of groups needed for a particular activity and practice ordinal numbers as they line up.

Teach students how to listen and respond appropriately to their peers. Students can practice with partners where one is speaking and the other is listening. One way to signal who is able to talk is by using a mouth on a stick and an ear on a stick. Give one child the mouth on a stick, which is the signal for that child to speak first, explaining their problem solving strategy first while the child holding the ear on a stick must only listen. Then they switch roles and the other child is able to explain. Students who are English Language Learners, those who receive speech services that would inhibit others' ability to understand them when they are speaking, or students who are non-verbal, can have the ear on a stick while their partner speaks. If they are able, these learners could then draw or label their paper rather than speak when it is their turn to share.

Practice with partners where one is speaking and the other is listening, and offer prompts for each role. The listener may use prompts such as, "Can you speak louder?" or "I didn't understand how you solved it. Can you explain it a different way?" or "Take me through the steps so I can see what you did." The speaker may use prompts such as, "I will go through each step in my process." Or "I am going to tell you how I solved the problem and the tools I used to help me reach a solution."

Teachers should model how students communicate one-on-one with a peer as well as how to communicate within a small group. It is not enough to simply tell your students to talk about how they found their answer. Teachers can build a community of problem solvers through the use of modeling, providing prompts and sentence starters as well as moving from simpler word problems to more complex problems throughout the school year.

1

Make Sense and Persevere

SMP 1—Make Sense of Problems and Persevere in Solving Them

How many minutes would a student have to work on a word problem in order for you to determine the student has persisted in solving the problem? Do students know what is meant by perseverance? How can students persist if they do not understand the context of the problem? Making sense of a problem begins with the ability to understand the words, symbols, numbers and figures in the problem. English Language Learners may not be familiar with regional dialect and outdated terms; early readers may be overwhelmed by a long paragraph; inexperienced problem solvers may not know how to interpret symbols and figures which accompany a problem.

What Does This Standard Mean for Grades K–2 Problem Solvers?

Students who are successful at SMP 1 read the problem more than once in order to make sense of the information provided. They are able to explain the problem to a peer in their own words. They can analyze what is given, what is not given and the goal of the problem. They can draw a picture or use objects to represent the situation in the problem. They can make several attempts to find the answer, considering the strategies of other problem solvers. They continually ask themselves if their strategy and their answer make sense.

In order to fully apply SMP 1 when approaching a word problem, students in Kindergarten through second grade should be able to take ownership of their procedures by using the following *I Can* statements:

- ◆ *I can explain the meaning of the problem in my own words.*
- ◆ *I can analyze what is given, what is not given and the goal of the problem.*
- ◆ *I can use a picture or concrete objects to understand and solve the problem.*
- ◆ *I can understand the strategies of others.*
- ◆ *I can ask myself, does this make sense?*

Seven Blind Mice by Ed Young

The book *Seven Blind Mice* (1992) represents a version of the folktale *The Blind Man and the Elephant* and follows seven blind mice who attempt to figure out the nature of a strange Something at the pond near their home. They must rely on their sense of touch to make sense of the object. Each mouse travels alone to the pond and feels a different part of the object, resulting in disagreements over the identity of the object. The last mouse to go to the pond, White Mouse, first listens to the ideas of the other mice concerning the identity of the object. She attempts a different strategy and runs across the strange Something from end to end, making sense of the various parts of the object mentioned by the other mice to reveal it is an elephant. You can use this book to illustrate SMP 1 by providing a concrete example of how to make sense of a problem and then persist at finding its solution as well as incorporate how to decompose the number 7.

Operations and Algebraic Thinking K.OA

Understand addition as putting together and adding to, and understand subtraction as taking apart and taking from.

Decompose numbers less than or equal to 10 into pairs in more than one way, e.g., by using objects or drawings, and record each decomposition by a drawing or equation.

As you read the book, engage students in a discussion by asking the following questions based on the *I Can* statements:

- ◆ What was the problem in the book and why was it a problem for them? *Explain the meaning of the problem.* (One day the mice found a strange Something at the pond and they were afraid because they didn't know what it was; they didn't know if the strange Something would drink all of the water at their pond or not let them get to the water.)

◆ What information could the mice use to solve their problem since they are blind? *Analyze what is given, what is not given and the goal.* (They could touch the strange Something but they couldn't see it; they could only use the information from their sense of touch to try to identify the object.)

◆ What were some of the ways the mice tried to solve their problem? *Use a picture or concrete objects to understand and solve the problem.* (They went one at a time to the pond to find the object; they each touched the object and then told the other mice what they thought it was; each mouse wanted to experience it for him/herself in order to figure out what it was.)

◆ What did White Mouse do differently than the other mice? *Ask yourself, does this make sense?* (She wasn't sure if the answers of the other mice made sense so she wanted to go to the pond too; she decided to touch every part of the object instead of just one part and see if she could agree with one of the answers.)

◆ How did she use what she learned from the other mice? *Understand the strategies of others.* (White Mouse listened to the ideas of the other mice before she went to the pond to find the object; she used descriptions given by the other mice to help form her answer.)

For a second reading of the book, model how to use the *I Can* statements to solve a word problem related to the story:

There were 7 blind mice who found a strange Something by their pond then ran back home. One of the mice, Red Mouse, went back to the pond the next day to touch the strange Something. While he was at the pond, how many mice were still at home?

Use a think-aloud method to begin modeling the *I Can* statements:

"First, I have to think, *can I explain the meaning of the problem in my own words?* There are 7 mice. One mouse is at the pond. How many mice are still at home? Next, I have to *analyze what is given, what is not given and my goal.* I am given the total number of mice at the beginning of the problem, which is 7. I am given the number of mice that went to the pond, which is 1. I am not given the number of mice that are still home, which is my goal. I can use 7 counters to *understand and solve the problem* by acting it out. I will use 7 counters as the mice at home. I will move 1 of the counters to an oval, which will be the pond. Now I can count the mice that are still at home. There are 6 mice still at home." (See Figure 1.1.)

Figure 1.1

Once the students have observed and listened to your think-aloud strategy, they can become more involved in applying the next *I Can* statement. Draw a picture based on your concrete model and ask the students if your picture represents the problem and if your answer makes sense while further modeling:

> "I'm not finished yet because I have to look back at my answer and ask myself, *does this answer make sense?* Well, if I started out with 7 mice and 1 mouse was taken away from the group, there would be 6 mice left because 6 is 1 less than 7. So my answer does make sense. Does anyone have another way, or strategy, to solve the problem?"

Students can share their ideas for solving the problem with a peer using objects, pictures, words and symbols based on their readiness levels and experiences. Ask students to share their work with the class and see if others came up with the same answer but used a different strategy. If students are able, they should explain their strategy and you can restate their explanation in order to model the last *I Can* statement:

> "Janet, let me see if I understand your strategy for solving the mouse problem. You drew 7 mice then you crossed 1 out because you said that would be the mouse at the pond. Then you counted the mice that were not crossed out and you also got 6 for your answer. Okay, I *understand your strategy for solving the problem.* Thank you, Janet."

It is important to do similar problems with students so they can generalize the procedures for making sense of a problem and trying different ways to solve it. Show the students a problem using a different number but a similar context:

> *There were 8 little rabbits who found a carrot patch but the farmer chased them back to their home. One of the rabbits, Brown Rabbit, went back to the patch to dig out a carrot. While he was at the carrot patch, how many rabbits were still at home?*

Help the students think about how they should approach this problem by asking some questions:

- ◆ How is this problem similar to the problem about the 7 blind mice? (There are animals that find something then run back home; one of the animals goes back while the others stay home; you have to find how many animals are still home.)
- ◆ How is it different? (There are rabbits instead of mice; there are 8 animals instead of 7; they find a carrot patch instead of a strange Something; they go home because the farmer chases them instead of being scared off by the strange Something; the rabbits are not blind.)

After the students have discussed similarities and differences between the two problems, use a think-aloud strategy to explain how to use the *I Can* statements with the rabbit problem:

"I can begin to solve this problem the way I solved the problem about the 7 blind mice. First, how *can I explain the meaning of the problem in my own words*? There are 8 rabbits. One rabbit is at the carrot patch. How many rabbits are still at home? Next, I have to *analyze what is given, what is not given and my goal*. I am given the total number of rabbits, which is 8. I am given the number of rabbits that went to the carrot patch, which is 1. I am not given the number of rabbits that are still home, which is my goal."

Before explaining the next step, solicit ideas from students to see if they can think of strategies to use to find the answer. You can ask guiding questions to help them:

- ◆ What can I do to find out how many rabbits are still at home?
- ◆ What strategy did I use with the problem about the 7 blind mice?
- ◆ Can I use that same strategy with this problem?
- ◆ Are there other strategies I could use?

Once they have shared ideas about how you could solve the problem, choose a different strategy so they can see there are other strategies that can work:

"Now I have to think about a strategy I can use to help me *understand and solve the problem*. For the 7 blind mice problem, I used counters and acted out the problem. That helped me see how many mice were still at home. This time I will use a different strategy. I'll use the strategy Janet shared with us. I will draw 8 circles, which will be the rabbits at the beginning of the problem. Then I will cross out 1 of the circles because that will be the rabbit that went to the carrot patch. I will draw 1 circle over here, which is the rabbit who left home. Now I can count the rabbits still at home. There are 7 rabbits still at home." (See Figure 1.2.)

MATH-terpieces: The Art of Problem Solving by Greg Tang

In the book *MATH-terpieces: The Art of Problem Solving* (2003), famous artwork serves as a backdrop for the reader to be able to add groups of objects to find sums up to a target number. Once you have modeled how to use the *I Can* statements using a few different word problems, students should practice using the *I Can* statements to solve problems with a peer as they use drawings and equations to add three whole numbers.

Operations and Algebraic Thinking 1.OA

Represent and solve problems involving addition and subtraction.

Solve word problems that call for addition of three whole numbers whose sum is less than or equal to 20, e.g., by using objects, drawings and equations with a symbol for the unknown number to represent the problem.

The first painting, *Ballet Rehearsal on Stage* by Edgar Degas, is displayed on the left side of the two-page spread with groups of ballet shoes on the right side.

The problem posed to the reader is:

Can you make 7 with these shoes?
Three clever ways earn rave reviews!

Figure 1.2

Ask questions that can allow students to solve the problem with their peers:

- Can you *explain the meaning of the problem in your own words* to your partner? (How can I make 7 by putting together groups of shoes? There are 3 ways to do it.)
- Can you *analyze what is given, what is not given and the goal* of this problem? (There are 5 different groups of ballet shoes in the picture. There is a group of 5, a group of 2, a group of 4, a group of 3 and 1 shoe by itself. I have all of the information I need to find the 3 ways to make 7.)
- Can you *use concrete objects or pictures to help you understand and solve the problem?* You can use paper to draw the shoes or cubes snapped together to represent the groups of shoes. Keep track of all of the combinations you and your partner try, even if some of your ideas don't work. (I can count groups of shoes to see if they add up to 7. I have solved the problem when I have found 3 different ways to make 7 with the groups of shoes in the picture.)
- What are your answers? *Do they make sense?* How can you check? (I can write number sentences for my answers and check with the cubes to make sure my 3 ways add up to 7.)
- What strategy did you and your partner use? Who would like to *explain their strategy* to the class? (We drew pictures of the shoes and the first one we wrote was 4 and 3 because we knew 4 plus 3 equals 7. Then we looked at the biggest group, which was 5, and we knew to put it with the 2 because we knew 5 plus 2 equals 7. Then we were stuck because there is 1 shoe left but there is no group of 6 shoes to go with it.)
- Did anyone else get only two ways to make 7? Who else got $4+3$ and $5+2$? Is there another way? Greg Tang said there are 3 ways to do it with the groups of shoes on the page. Do you have to use only 2 groups of shoes in your answers? Can you use a group of shoes more than once? Can you do it with 3 groups of shoes? (Yes. We found the last way. You have to use 3 groups of shoes. You can use the group of 2 shoes, the group of 4 shoes and the 1 shoe that was leftover. We know this works because we checked and $2+4+1=7$.)

If the students had difficulty finding the solution with three addends, use a think-aloud method to explain it:

"When I was trying to solve the problem, I tried to use a system. I began with the largest group of shoes first, the group of 5, and looked for a number I could add to 5 to equal 7. I know $5+2=7$ so I used the group of 2 shoes. Then I used the second largest group of shoes, which was 4, and I matched that with the group of 3 shoes because I know $4+3=7$. The only shoe left was 1 by itself and there was no group of 6 shoes to add to it to equal 7, since I know $6+1=7$. But then I thought there might be a way to combine two groups of shoes to make 6. If I used the group of 4 shoes and the group of 2 shoes I could make a group of 6 and that would go with the 1 shoe by itself."

These examples illustrate not only the use of the five *I Can* statements for SMP 1 but offer opportunities for children to persist in finding the less obvious way to find the sum of 7 in the problem, using three addends rather than just two. Some partners may have found all three ways to make 7 with little effort. Be sure to allow enough time for students to try to discover all three ways with their partner as well as time for children to share their strategies with the class. Some may use drawings, some can add cubes or other manipulatives, and some may be at the abstract level and only use the number of shoes in each group to find the sums.

Often, when students realize some of their classmates have already found the solution, they are motivated to persist and find the solution themselves. They realize a solution does exist and if their peers can find it, so can they. Other students may actually feel discouraged when their classmates find the solution quickly while they are still struggling to find it. Give students many opportunities to discuss ideas and strategies with their peers so they can see there are many ways to solve a problem. Switch up partners often so students can work with peers of various levels of mathematical competency and understanding.

Read other pages in the *MATH-terpieces* book so students can practice solving the problems with their partner and start using strategies shared by their peers. As they work with their partner on a few more problems, ask guiding questions such as:

◆ How would you explain the meaning of this problem to another person?
◆ What information was given and what was your goal?
◆ Did you use the same strategy as on the first problem? Which strategy did you use?
◆ Which problems were easy to solve? Which problems were tricky? Why?
◆ Did you have to add more than two groups of objects like we did in the first problem? Give me an example where you did this.

◆ How did you check to see if you answers made sense?
◆ Did you try strategies suggested by your classmates? Which ones?

Splash! by Ann Jonas

In the book *Splash!* (1995) a girl tells a story about her many pets and how one day some of them fell or jumped into the pond while others came out of the pond. At the bottom of each page the same question is repeated, "How many are in my pond?" Students who have had practice using the five *I Can* statements while solving problems with their peers may be ready to try solving some on their own with addition and subtraction.

Operations and Algebraic Thinking 2.OA

Represent and solve problems involving addition and subtraction.
Use addition and subtraction within 100 to solve one- and two-step word problems involving situations of adding to, taking from, putting together, taking apart and comparing, with unknowns in all positions, e.g., by using drawings and equations with a symbol for the unknown number to represent the problem.

At first, provide a checklist with the five statements so students can begin to learn them and eventually use them automatically. Provide a large piece of paper and a manipulative such as circle counters so they can draw a picture or use the concrete objects to solve the problem. Challenge students by telling them you have a longer problem for them to solve based on a book you will soon read. You will give them time to try solving it on their own. After about ten minutes, you will all discuss the problem and then you will give them more time to go back and try again to solve it on their own if they need it. Display the following problem:

I have a pond in my backyard.
I have one turtle, two catfish, three frogs, and four goldfish.
All of my fish are in the pond.
The turtle jumps into the pond.
One frog jumps in. My cat falls in. My dog falls in.
How many are in my pond?

As students are working, walk around to see what they are drawing on their paper or modeling with their counters. Encourage students to think of as many different strategies as they can to find the answer. Let them know they should not say the answer if they think they have found it but just write it on the top of their paper and circle it. After most students have attempted to solve the problem, stop them and ask guiding questions for the discussion:

- Can you *explain the meaning of the problem in your own words*?
- What *information is given, what is not given and what is your goal*?
- What strategies did you use when you started to solve the problem? Did you *use a picture or concrete objects to understand and solve the problem*?
- Why did you choose that strategy?

If there were many who did not get the correct answer, did not finish solving the problem or were having difficulty choosing an appropriate strategy, allow students to talk with their peers about what they have drawn or represented, then provide 5–10 more minutes to work on the problem. Walk around again to see if they are using the information from their discussion to solve the problem. When most have provided an answer, stop and ask:

- Did anyone try to *understand the strategies of others* in our class? Which strategies?
- How did you *check your answer to see if it made sense*?

Then read the book and talk about how the illustrations can provide the answers to the question on each page because we can count the animals in the pond. Let them know this is another strategy to solve the problem. As students become independent problem solvers, provide them with questions they can use to help them make sense of a problem, perhaps in the form of a poster hung in the class:

- What is the setting or situation in the problem?
- Would I be able to draw a picture to show what is happening in the problem?
- Do I understand the words, symbols and numbers in the problem?
- Are there any words I don't know? Are there illustrations or other words in the problem I can use to help me figure out unknown words?
- Can I use a strategy I used on another problem?
- Can I use a strategy suggested by another student?

What Does This Standard Mean for Grades 3–5 Problem Solvers?

Students in intermediate grades who are successful at SMP 1 can restate the problem in their own words in order to make sense of the information provided. They can establish a plan to reach a solution and explain their plan to a peer. They can analyze what is given, what is not given and the goal of the problem. They can talk about their problem solving process as well as using a drawing, a table or a graph to represent the situation in the problem. They continually ask themselves if their strategy and their answer make sense, using a different method to attempt to reach the same answer.

In order to fully apply SMP 1 when approaching a word problem, students in intermediate grades should be able to apply the following *I Can* statements:

- ◆ *I can restate the meaning of the problem and create a plan to reach a solution.*
- ◆ *I can analyze what is given, what is not given and the goal of the problem.*
- ◆ *I can use verbal descriptions, tables, graphs and diagrams to solve the problem.*
- ◆ *I can determine whether or not the answer makes sense by using a different method.*

The Warlord's Kites by Virginia Walton Pilegard

In the book *The Warlord's Kites* (2004) the warlord's son, Chuan, and the puppet master's daughter, Jing Jing, work together to create kites that look like ghosts in order to save the palace from invasion. As they work on the kites they use their hands to measure a specified length and width for each kite. Students can use the illustrations along with Chuan and Jing Jing's explanation of how to calculate the area of a rectangle.

Measurement and Data 3.MD

Geometric measurement: understand concepts of area and relate area to multiplication and to addition.

Find the area of a rectangle with whole-number side lengths by tiling it, and show that the area is the same as would be found by multiplying the side lengths.

As you read the book, engage students in a discussion by asking the following questions based on the *I Can* statements:

◆ What was the problem in the book? *Restate the meaning of the problem and create a plan to reach a solution.* (An army was invading the warlord's palace; Jing Jing said she could save the palace by making 3 square kites.)

◆ What information do Chuan and Jing Jing need to make the kites? Is there information they do not have? What is their goal? *Analyze what is given, what is not given and the goal of the problem.* (They need to measure the area of the paper to make a square; they do not have a ruler to measure the paper; their goal is to make 3 square kites.)

◆ What did Chuan and Jing Jing do to solve their problem? *Use verbal descriptions, tables, graphs and diagrams to solve the problem.* (Jing Jing said they need the paper to be cut 6 hands high and 6 hands wide; Chuan used his hands to measure the area of the paper by printing 36 hands on the paper in order to create a 6 by 6 square.)

◆ Why did Chuan make 36 hand prints on the first piece of paper? What did he state as the "formula"? What other method did Chuan, Jing Jing and the warlord's son use instead of making all 36 hand prints on the next 2 kites? *Determine whether or not the answer makes sense by using a different method.* (Chuan made all 36 hand prints on the first piece of paper because he wanted to show Jing Jing that it made a square; Chuan's formula is to multiply length times width; their other method was to only make 6 hand prints for the length and 6 for the width, which is 36 if you multiply length times width.)

For an extended activity, practice using the *I Can* statements to solve a word problem related to the story:

Chuan and Jing Jing want to make a new kite but they didn't want it to be a square. They want to make it longer on one side so it will look like a rectangle. They want the total area to be 40 hands. What length and width could they use for the rectangle?

Prompt the students to use the *I Can* statements to solve the problem:

"First, we have to think, *can I restate the meaning of the problem in my own words?* Turn to your partner and use your own words to explain the meaning of the problem. Next, we have to *analyze what is given, what is not given and the goal of the problem.* Talk to your partner about the following: What information about the measurements of the rectangle is given in the problem? What information about the measurements of the rectangle is not given in the problem? According to the problem, what is the goal of your new problem? Now take about five minutes to draw a diagram of the new kite and then share with your partner by *using verbal descriptions*, along with your *diagram to solve the problem.*"

◆ You can have square tiles or graph paper available for students who may need to create a model first. Some students may benefit from reviewing the factors of 40 in order to work backward to apply the formula of multiplying length times width. Remind students to use only whole numbers at this point in time.
◆ Ask students to label the length and width of their diagram and then post all of the possible dimensions of the kite so the class can see that the length and width should always be 40 because that is the area of the desired kite in the problem.
◆ If this problem is too difficult for some students, use smaller numbers, such as an area of 20 instead of 40. You could also do this problem together as a class and create a similar problem for students to solve with a partner or independently.

A Remainder of One by Elinor J. Pinczes

In the book *A Remainder of One* (1995), a troop of 25 insect soldiers are marching in a parade for their Queen. The Queen likes things tidy so she is not pleased when she sees a soldier, Joe, without a partner when the troops are divided into two lines. Joe is removed from the parade and that night he thinks that if the troop divides into three lines, he might not be alone at the end of the pack. He continues to think of how the troop of 25 can be divided into other configurations until finally he is not left alone at the end of the pack.

Operations and Algebraic Thinking 4.OA

Use the four operations with whole numbers to solve problems.

Solve multi-step word problems posed with whole numbers and having whole-number answers using the four operations, including problems in which remainders must be interpreted. Represent these problems using equations with a letter standing for the unknown quantity.

As you read the book, engage students in a discussion by asking the following questions based on the *I Can* statements:

◆ At the beginning of the story, what is the problem with the squadron of 25 insects marching in the parade? How can they solve it? *Restate the meaning of the problem and create a plan to reach a solution.* (The insect soldiers are divided into two lines and each one has a partner except for the last soldier, Joe, who is marching by himself at the back; they have to figure out a configuration so Joe is not left out of the pack.)

◆ What information does Joe have to figure out how he can be included? Is there information he does not have? What is his goal? *Analyze what is given, what is not given and the goal of the problem.* (Joe knows the Queen likes things tidy so she doesn't want a soldier by himself at the end of the pack; he does not know how the Sergeant will divide up the troops the next day; his goal is to figure out if there is another way to divide the 25 insects into lines of soldiers so that he won't be left alone.)

◆ What did Joe think about at night to solve his problem? How can we keep track of the configurations of the 25 insects? *Use verbal descriptions, tables, graphs and diagrams to solve the problem.* (Joe thought maybe one more line would make everything right so he won't be the "remainder of one"; we can make a table to keep track of the configurations.)

◆ So our multi-step problem could be:

> There is a troop of 25 soldiers marching in a pack. What are all of the configurations they can create for their pack? Which configuration would allow the same number of soldiers in each line? Show each of your attempts, justify the correct answer and explain why the other attempts do NOT work.

◆ Let's make an equation, work out the division problem and draw a diagram for the first configuration to see why it didn't work. (See Table 1.1.) Our equation is 25 divided by 2 because there are 25 soldiers in the troop and the book states that each troop

Table 1.1

Attempt at Dividing Troop	Division with Area Model	Diagram to Interpret Remainder
$25 \div 2 = n$	$\begin{array}{r} 12 \\ 2\overline{\smash{)}25} \\ -24 \\ \hline 1 \end{array}$	

divided by 2 in the parade. We will use n for the unknown quantity in our equation.

◆ Now we will use the area model to perform the division operation. Our first step is to think about our multiplication facts for 2. Can we multiply 2 by a whole number so the product would be 25? Why or why not? (No, because multiplying by 2 always results in an even number and 25 is an odd number; no, because if we list out multiples of 2 we can see it will not include 25.) Since we cannot get a product of 25, how close can we get to a product of 25? (We can get a product of 24 by multiplying 2 by 12.)

◆ Now let's look at our area model. We put 12 at the top of the box because 12 is the factor we multiplied by 2 to get the product of 24. The next step is to subtract 24 from 25 and we see there is a remainder of 1. What does the remainder represent in our word problem? (It represents the soldier that does not fit into the configuration.)

◆ Now we can draw a diagram in our table to represent our configuration so we can see how the remainder represents the lone soldier at the end of the pack. *Determine whether or not the answer makes sense by using a different method.*

◆ On the next day, we see the troop has divided by 3. Let's make an equation, work out the division problem and draw a diagram for the second configuration to see why it didn't work. (See Table 1.2.) Our next equation is 25 divided by 3 because there are 25 soldiers in the troop and the book states that each troop divided by 3 in the parade. We will use n for the unknown quantity in our equation.

◆ Now we will use the area model to perform the division operation. Our first step is to think about our multiplication facts for 3. Can we multiply 3 by a whole number so the product would be 25? Why or why not? (No, because if we list out multiples of 3 we can see that we will not include 25.) Since we cannot get a product of 25, how close can we get to a product of 25? (We can get a product of 24 by multiplying 3 by 8.)

Table 1.2

Attempt at Dividing Troop	Division with Area Model	Diagram to Interpret Remainder
$25 \div 3 = n$	$$\begin{array}{r} 8 \\ 3\,\overline{\smash{)}\,25} \\ -24 \\ \hline 1 \end{array}$$	

◆ Now let's look at our area model. We put 8 at the top of the box because 8 is the factor we multiplied by 3 to get the product of 24. The next step is to subtract 24 from 25 and we see there is a remainder of 1. What does the remainder represent in our word problem? (It represents the soldier that does not fit into the configuration.)

◆ Now we can draw a diagram in our table to represent our configuration so we can see how the remainder represents the lone soldier at the end of the pack. *Determine whether or not the answer makes sense by using a different method.*

◆ On the next day, we see the troop has divided by 4. Let's make an equation, work out the division problem and draw a diagram for the third configuration to see why it didn't work. (See Table 1.3.) Our next equation is 25 divided by 4 because there are 25 soldiers in the troop and the book states that each troop divided by 4 in the parade. We will use n for the unknown quantity in our equation.

◆ Now we will use the area model to perform the division operation. Our first step is to think about our multiplication facts for 4. Can we multiply 4 by a whole number so the product would be 25? Why or why not? (No, because if we list out multiples of 4 we can see it will not include 25; no, because multiplying by 4 always results in an even number and 25 is an odd number.) Since we cannot get a product of 25, how close can we get to a product of 25? (We can get a product of 24 by multiplying 4 by 6.)

◆ Now let's look at our area model. We put 6 at the top of the box because 6 is the factor we multiplied by 4 to get the product of 24. The next step is to subtract 24 from 25 and we see there is a remainder of 1. What does the remainder represent in our word problem? (It represents the soldier that does not fit into the configuration.)

◆ Now we can draw a diagram to represent our configuration so we can see how the remainder represents the lone soldier at the end of the pack. *Determine whether or not the answer makes sense by using a different method.*

Table 1.3

Attempt at Dividing Troop	Division with Area Model	Diagram to Interpret Remainder
$25 \div 4 = n$	$\begin{array}{r} 6 \\ 4\ \boxed{25} \\ -24 \\ \hline 1 \end{array}$	

Table 1.4

Attempt at Dividing Troop	Division with Area Model	Diagram to Interpret Remainder
$25 \div 5 = n$	$\begin{array}{r} 5 \\ 5\ \boxed{25} \\ -25 \\ \hline 0 \end{array}$	●●●●● ●●●●● ●●●●● ●●●●● ●●●●●

- On the last day, we see the troop has divided by 5. Let's make an equation, work out the division problem and draw a diagram for the last configuration to see why it worked. (See Table 1.4.) Our next equation is 25 divided by 5 because there are 25 soldiers in the troop and the book states that each troop divided by 5 in the parade. We will use *n* for the unknown quantity in our equation.
- Now we will use the area model to perform the division operation. Our first step is to think about our multiplication facts for 5. Can we multiply 5 by a whole number so the product would be 25? Why or why not? (Yes, because multiples of 5 have a 5 in the ones place; when we list out multiples of 5 we see it will include 25.)
- Now let's look at our area model. We put 5 at the top of the box because 5 is the factor we multiplied by 5 to get the product of 25. The next step is to subtract 25 from 25 and we see there is no remainder. Why don't we have a remainder? What does this represent in our word problem? (We don't have a remainder because 25 is evenly divided by 5: it represents the fact that the soldier does fit into the configuration so there are no soldiers left alone in the back of the pack.)
- Now we can draw a diagram to represent our configuration so we can see why there is no remainder. *Determine whether or not the answer makes sense by using a different method.*

Multiplying Menace by Pam Calvert

The book *Multiplying Menace: The Revenge of Rumpelstiltskin* (2006) is a book about the return of Rumpelstiltskin, who has come for the Queen's ten-year-old son, Peter. When she refuses to give up her son, Rumpelstiltskin uses his walking stick to make people and animals disappear while creating hundreds of insects and other pests around the kingdom. Peter agrees to go and notices that Rumpelstiltskin points the stick at something and yells out a command such as, "Candle times eight." Peter steals the stick and is able to reduce the number of pests by shouting a command that involves multiplying by a fraction.

> **Number and Operations—Fractions 5.NF**
>
> **Apply and extend previous understandings of multiplication and division to multiply and divide fractions.**
>
> Apply and extend previous understandings of multiplication to multiply a fraction or whole number by a fraction.

As you read the book, engage students in a discussion by asking the following questions based on the *I Can* statements:

◆ At the beginning of the story, what is the problem after Rumpelstiltskin leaves the Queen? How can they solve it? *Restate the meaning of the problem and create a plan to reach a solution.* (Rumpelstiltskin made guards and farm animals disappear, put holes in the castle walls, gave the King six noses and put hundreds of rats and insects around the kingdom; they have to figure out how to undo Rumpelstiltskin's tricks and make sure he is gone forever.)

◆ What information does Peter find out when he goes with Rumpelstiltskin? Is there information he does not have? What is his goal? *Analyze what is given, what is not given and the goal of the problem.* (Peter knows that Rumpelstiltskin uses his stick to make things change; he does not know how the stick works; his goal is to figure out how the stick works so he can steal it and use it to change things back to the way they were.)

◆ How can we keep track of Rumpelstiltskin's tricks so we can figure out how Peter can solve his problem? *Use verbal descriptions, tables, graphs and diagrams to solve the problem.* (We can use a table to keep track of the multiplication commands in the story and then use the pictures or words in the story to figure out the total number of objects.)

◆ Let's write out the commands that Rumpelstiltskin gives when he first arrives at the kingdom and when he is in his hut with Peter and see if we can figure out each total with the illustrations and words in the story. (See Table 1.5.)

◆ His first command is "stone times one-third" and we can see some of the stones in the castle wall have been reduced to one-third of their original length. His second command is "nose times six" and we can see he gives the King six noses. Once he and Peter are at the hut, his command is "candle times eight" and we can see eight

Table 1.5

Command	Total
Stone times one-third	One-third of a stone
Nose times six	Six noses
Candle times eight	Eight candles
Branches times ten	Forty branches
Pies times five	Ten pies

candles on the table. Then Rumpelstiltskin puts four branches in the fireplace and says "branches times ten" and the book states 40 branches appeared. There are two meat pies on the table and he says "pies times five" and the book states ten pies appeared.

◆ Now that we have used verbal descriptions and pictures to keep track of Rumpelstiltskin's tricks, we have to figure out the secret of the stick so Peter can solve his problem. Let's make another column in our table to create an equation for each command. (See Table 1.6.) *Determine whether or not the answer makes sense by using a different method.*

◆ Let's write out the commands that Peter gives when he steals the stick and see if we can figure out what he has to do with the stick to change things back to the way they were. (See Table 1.7.)

◆ We see Peter point the stick at the four remaining pies on the table and he says "pie times five" but nothing happens. What does he figure out about how the stick works? (He has to use the end of the stick with the multiplication symbol when he points to the objects and says the command.) Then he points the end of the stick with the **X** carved into it at the four pies and repeats the command, "pie times five," and then 20 pies appear on the table.

Table 1.6

Command	Total	Equation
Stone times one-third	One-third of a stone	$1 \times 1/3 = 1/3$
Nose times six	Six noses	$1 \times 6 = 6$
Candle times eight	Eight candles	$1 \times 8 = 8$
Branches times ten	Forty branches	$4 \times 10 = 40$
Pies times five	Ten pies	$2 \times 5 = 10$

Table 1.7

Peter's Command	Total	Equation
Pies times five	Twenty pies	$4 \times 5 = 20$
Pies times one-third	Four pies	$12 \times 1/3 = 4$
Chairs times fifty	One hundred chairs	$2 \times 50 = 100$
Stepping stones times nine	Twenty-seven stones	$3 \times 9 = 27$
Grasshoppers times one one-hundredth	One grasshopper	$100 \times 1/100 = 1$
Nose times one-sixth	One nose	$6 \times 1/6 = 1$
Rumpelstiltskin times zero	Zero Rumpelstiltskin	$1 \times 0 = 0$

◆ But Peter realizes that Rumpelstiltskin will know he switched the magic stick with a broom handle if he sees that the number of pies has increased. He eats eight pies but when he can't eat any more, what number does he use so the number of pies will decrease and why? (He uses one-third because one-third of 12 is 4; if he multiplies a whole number by a fraction it will reduce the total and he wants the total to be 4 again.)

◆ Before he leaves the hut, what does Peter do to prevent Rumpelstiltskin from getting out of his bed? (He says a command so that the 2 chairs turn into 100 chairs.)

◆ When he meets the boy stranded on a stone in the river, how does Peter help him? (He points the stick at three stones and says "stepping stones times nine.")

◆ When he sees the grasshoppers eating the plants in the royal garden, what number does he use so the number of insects will decrease and why? (He uses one one-hundredth because multiplying 100 by 1/100 he would change the 100 grasshoppers into 1 grasshopper; multiplying a number by its inverse results in a product of 1.)

◆ What number does he use so the King can have one nose again and why? (He uses one-sixth because multiplying 6 by 1/6 he would change the King's 6 noses into 1 nose; multiplying a number by its inverse results in a product of 1.)

◆ When Rumpelstiltskin shows up at the palace at the end, what number does Peter use so he will disappear and why? (He uses 0 because multiplying any number by 0 will result in a product of 0 and Rumpelstiltskin will be gone.)

Wrapping It Up

Students have to learn to be patient in their quest for perseverance. They should learn it is okay to put aside a problem and come back to it later, like they did in the problem from the book *Splash!*, or to use a different method to try to reach the same answer, as they did in the area problem in *The Warlord's Kite* or the division problem in *A Remainder of One*. Often a solution emerges when we represent a problem through a combination of methods, such as diagrams, pictures, verbal descriptions and equations, or we hear how others are thinking about solving the same problem.

2

Reason Abstractly
SMP 2—Reason Abstractly and Quantitatively

Students are exposed to problem solving in Kindergarten in a concrete way, with physical objects such as toys or blocks, or by acting out a problem solving situation. When students are in first and second grade, they gradually move away from the concrete stage of learning and are exposed to number models representing the problem solving situations. Exposing students to this connecting stage, in which there are still physical or visual representations alongside the numbers and symbols, is critical so students will be able to successfully reach the abstract stage in third through fifth grade.

What Does This Standard Mean for Grades K–2 Problem Solvers?

Students who are successful at SMP 2 can see relationships among the numbers so they can make decisions about which operation would be most efficient to solve the problem. They can decontextualize the situation by representing it symbolically ($5 + ? = 10$), but they must also contextualize the situation by examining the referents (objects, people, etc.) in the problem in order to determine which symbols to use. Through experience with various situations, students can determine whether the situation involves a group added to another group, if there are groups being compared, if there are groups taken from a larger group, as well as which operation is represented by these actions. It is also crucial for problem solvers to be able to recognize and use the properties of operations when applying SMP 2.

In order to fully apply SMP 2 when approaching a word problem, students should be able to take ownership of their procedures by using the following *I Can* statements:

◆ *I can understand how the numbers in the problem are related.*
◆ *I can use the units in the problem.*
◆ *I can use properties of operations.*
◆ *I can represent the problem with symbols.*

Each Orange Had 8 Slices by Paul Giganti, Jr.

There are opportunities in *Each Orange Had 8 Slices* (1992) for students to count familiar objects such as flowers, cows, trucks and, of course, orange slices. Each page illustrates a group of objects with an equal number of smaller objects with the question, "How many?" for the reader to answer. Students can start to use units in their answer and can make connections between counting and addition.

Operations and Algebraic Thinking K.OA

Count to tell the number of objects.

Count to answer "how many?" questions about as many as 20 things arranged in a line, a rectangular array, or a circle, or as many as ten things in a scattered configuration; given a number from 1–20, count out that many objects.

As you read the book, use a think-aloud method to demonstrate how to use the *I Can* statements:

"On the first page there are 3 red flowers. On each flower there are 6 petals. There is a question for me to answer: How many red flowers were there? I can find the answer to this question by counting the flowers or by reading the words on the page. It says there are 3 red flowers. There is another question: How many pretty petals were there? This question is asking about the total number of petals on the page. But first I have to go back to the number of flowers, which is 3. I know there are 6 petals on each flower because it says there are 6. I have to think about *how the numbers in the problem are related*. There are 6 petals on each flower and there are 3 flowers. So I will have to count to 6, then keep counting the petals on the second flower, which brings me up to 12, and then continue counting the petals on the last flower. When I am solving a word problem I have to *use the units in the problem* so the answer is, there are 18 petals."

If your students are ready to use the *I Can* statements, lead them into an activity using the tiny black bugs on the page:

◆ Let's review our answers for the first two questions: How many red flowers were there? (There were 3 flowers.) How many pretty petals were there? (There were 18 petals.)

◆ There is one more question on this page: How many tiny black bugs were there in all? This page says that each petal had 2 tiny black bugs and we can see there are 2 black bugs on each petal. When we are counting objects that are in a circle, we have to keep track of where we started counting, so we don't count an object more than once.

◆ Let's start counting the black bugs on the first flower, the flower on the left side of the page. I am going to put my left thumb on the petal at the top of the flower, right in the middle, so we know where we started counting the black bugs. (Count the bugs with the class, pointing to each bug until all are counted.) The answer is 36. But I can't forget to *use the units in the problem.* I will read the last question on this page and everyone can say the answer with the units, "How many tiny black bugs are there in all?" (There are 36 tiny black bugs.)

◆ On the next page we see 3 children and each child is riding a tricycle. How many wheels are on a tricycle? (There are 3 wheels.) Yes, the word tricycle has the same beginning as the word triangle, because "tri" means 3. Let's answer the questions on the page and *use the units in the problem* in our answer. How many kids were there? (There were 3 kids.) How many tricycles were there? (There were 3 tricycles.)

◆ Just like the last question on the first page, we may have to do some counting to find the answer to the problem. We will have to think about *how the numbers in the problem are related.* The question is, "How many wheels were there in all?" How can we find the answer to this problem? (We can count all of the wheels on the page.)

◆ Let's start with the red tricycle on the left side of the page. We already know there are 3 wheels so instead of counting "1, 2, 3," let's all say "3" and then continue counting with the wheels on the green bike. We will have to think about what comes after "3" when we count. Okay, here we go. (Point to the red tricycle and say "3," then point to the first wheel on the green tricycle and continue counting, "4, 5, 6," then point to the first wheel on the blue tricycle and continue counting, "7, 8, 9.") I will read the last question on this page and everyone can say the answer with the units, "How many wheels were there in all?" (There were 9 wheels.)

Continue counting objects on each page with your students, reinforcing the units in the answer to each question. If your students are ready to move from counting to adding, they can work on the last "in all" question on each page with a partner, using hands-on materials to represent the objects on the page. Or before you show students a page, read the first sentence and see if your students can count out that number of objects at their seat.

Ten Flashing Fireflies by Philomen Sturges

It is important in the early grades to expose students to the Commutative Property of Addition and to use this term, rather than refer to addition facts that follow this property as *turnaround facts, flip-flop facts,* or other sim-plified terms. In the book *Ten Flashing Fireflies* (1995), a young boy and girl are outside catching fireflies in a jar. As the children catch the fireflies one by one, the fireflies captured in the jar are shown on the left and the remaining fireflies in the sky are shown on the right of the two-page spread. This provides an opportunity to explore the Commutative Prop-erty of Addition as well as practice the addition facts for 10.

Operations and Algebraic Thinking 1.OA

Understand and apply properties of operations and the relationship between addition and subtraction.

Apply properties of operations as strategies to add and subtract. *Examples: If $8+3=11$ is known, then $3+8=11$ is also known (Commutative Property of Addition).*

As you read the book, students can keep track of the number of fireflies in the jar and in the sky by creating a chart to organize the information. (See Table 2.1.)

After reading the story, ask the class some questions about the informa-tion in the table:

- What do you notice about the numbers in each row? (They add up to 10.)
- What do you notice about the numbers in the first column? (They are in counting order from 1 to 10).
- What do you notice about the numbers in the second column? (They are in backward counting order from 10 to 0).
- Let's look at the column labeled Equations. How is each equation created from the numbers in that row? (The equation is the first number plus the second number and it equals 10.)

Table 2.1

Fireflies in the Jar	Fireflies in the Sky	Total Number of Fireflies	Equations
0	10	10	$0+10=10$
1	9	10	$1+9=10$
2	8	10	$2+8=10$
3	7	10	$3+7=10$
4	6	10	$4+6=10$
5	5	10	$5+5=10$
6	4	10	$6+4=10$
7	3	10	$7+3=10$
8	2	10	$8+2=10$
9	1	10	$9+1=10$
10	0	10	$10+0=10$

◆ Let's write the first equation and the last equation next to each other:

$$0+10=10, 10+0=10$$

◆ Do you notice any similarities or differences in these two equations? (They both have the same numbers—0 and 10. They both equal 10. They have the numbers in a different order. One has 0 first then 10. The other one has 10 first then 0.)

◆ Let's look for this relationship in other sets of equations in this column. I want you to write down as many as you can find. (Call on students to share their answers, asking each child to tell you how the two number models are similar and how they are different, based on your example.)

◆ There is a special name for math facts that use the same two addends but in a different order, such as these. It is called the Commutative Property of Addition. It makes it easier to learn your math facts because if you know one fact, such as $8+2=10$, then you know that $2+8$ will also equal 10.

Rooster's Off to See the World by Eric Carle

After practicing the Commutative Property of Addition, your students will be ready to explore the Associative Property of Addition with the book *Rooster's Off to See the World* (1972). In this book, one rooster sets off to see the world and encounters several animals on his journey. There is a pattern to the number of animals who join the rooster; in the top right corner of each page a small graphic is shown representing each set of animals as they appear in the story, and then each set of graphics disappears as the animals change their mind and return home. You can also review the Commutative Property before you introduce the Associative Property.

Operations and Algebraic Thinking 1.OA

Understand and apply properties of operations and the relationship between addition and subtraction.

Apply properties of operations as strategies to add and subtract. *Examples: If $8+3=11$ is known, then $3+8=11$ is also known (Commutative Property of Addition). To add $2+6+4$, the second two numbers can be added to make a ten, so $2+6+4=2+10=12$ (Associative Property of Addition).*

As you read the book, create a similar table to keep track of the number of animals who join the rooster. (See Table 2.2.)

After reading the story, ask students some questions about the information in the table:

◆ How did the number of animals change in each row? (There was only 1 animal in the first row. Then there were 2 more animals, then 3 more animals, then 4 more animals then 5 more animals.)

Table 2.2 **Visual**

Rooster	Cats	Frogs	Turtles	Fish	Equations
1 rooster					$1+0=1$
1 rooster	2 cats				$1+2=3$
1 rooster	2 cats	3 frogs			$1+2+3=6$
1 rooster	2 cats	3 frogs	4 turtles		$1+2+3+4=10$
1 rooster	2 cats	3 frogs	4 turtles	5 fish	$1+2+3+4+5=15$

- ◆ How did our equation change in each row? (First there was only 1 animal then there were 3, then 6, then 10, then 15. We had to keep adding another number but we started with 1 each time.)
- ◆ Let's look at the first equation, $1+0=1$. What is $0+1$? How do you know? (It is also 1. It is the same answer as $1+0$.) Do you remember what this property is called, when we change the order of the two addends but we get the same answer? (Commutative Property of Addition.)
- ◆ Can we use the Commutative Property with the second equation? (Yes. We can write $2+1=3$.)

Pass out 15 squares to each student, with a picture of each animal on a square, corresponding to the number of each animal in the story. Have the students use the square with the rooster, the two squares with a cat on each and the three squares with a frog on each.

- ◆ Let's use our pictures of animals to represent the third equation, $1+2+3=6$. We are going to add the first two animals, 1 rooster and 2 cats. How many animals do we have? (We have 3 animals.)
- ◆ Now we will add the 3 frogs. How many animals do we have now? (We have 6 animals.)
- ◆ Let's try it another way. Let's start with the 2 cats and add the 3 frogs. How many animals do we have? (We have 5 animals.) Now we will add the 1 rooster. How many animals do we have now? (We have 6 animals.)
- ◆ Did it matter which order we added the animals? Would we get the same answer if we started with the 1 rooster, added the 3 frogs and then added the 2 cats? Let's try it.
- ◆ So we found out that we will get the same answer if we add the numbers in a different order. This is called the Associative Property of Addition. You can use this property when there are more than two addends because you might find it easier to add them in a different order.

If your students are able to apply this property with three addends, use the rest of the table to have them make new equations with four and five addends. Encourage them to find an easier way to add multiple addends, such as making a ten or starting with the larger number.

Spaghetti and Meatballs for All! by Marilyn Burns

Students should also practice using SMP 2 for geometry concepts with the book *Spaghetti and Meatballs for All!* (1997). In this book, Mr. and Mrs. Comfort are planning a dinner party for a total of 32 people. They have to rent tables and chairs in order to seat everyone but as soon as guests arrive, the tables start getting pushed together so people can sit together. Students can explore the concept of how to form rows and columns as they use squares to make rectangles for the various seating configurations.

Geometry 2.G

Reason with shapes and their attributes.

Partition a rectangle into rows and columns of same-size squares and count to find the total number of them.

As you read the book, use a think-aloud method to demonstrate how to use the *I Can* statements:

> "Let's keep track of the number of people who arrive at the dinner party. So far, there is Mr. and Mrs. Comfort, their daughter, her husband and their 2 children. That is 6 people. When they arrive, they want to push some of the square tables together so they can sit with each other. Mrs. Comfort begins to say, 'But there won't be room…' So the problem I have to figure out is, will there be enough room for 6 people to sit at the 2 tables after they are pushed together? When I look at the illustration in the book I see there are 4 chairs around each table. That means that 8 people could sit at those 2 tables. I'm going to *represent the problem with symbols*. (See Figure 2.1.) The squares are tables and the little rectangles are the seats."

> "Now I have to determine how many people will be able to sit around those 2 tables after they push them together. (See Figure 2.2.) Again, I'm going to *represent the problem with symbols*."

Figure 2.1

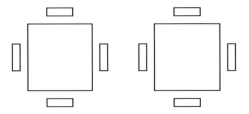

Figure 2.2

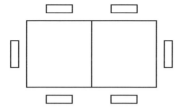

"I can see when they pushed those 2 tables together they had to remove the 2 chairs on the inside. When I represent this in my drawing, I can see there are only 6 chairs so that means 6 people can sit around the tables. I can use my drawing and I can *use the units in the problem* to think about how to represent my answer. In this problem there are several units—tables, chairs and people. There are 2 tables and there are 6 chairs around the tables. I know 1 person sits at each chair so there are 6 people that can sit around the tables. The answer to this problem should use the unit, *people*. There will be enough room for 6 people to sit at the 2 tables after they are pushed together."

Continue reading the book and let students draw a picture to help figure out the next problem:

◆ On the next page, 6 more guests arrive and someone suggests pushing 2 more tables over to the 2 tables already pushed together. Mrs. Comfort says, "But that won't work." Let's see what she means by that. How many people are at the dinner party now? (There were 6 people and 6 more people arrived; 6 + 6 is 12 so there are 12 people there now.)

◆ Our new problem is to see if there will be enough room for 12 people to sit at the 4 tables after they are pushed together. How can you *use symbols to represent the problem*? (We can draw symbols for the tables and chairs.) Yes, we can draw the 4 tables pushed together. (See Figure 2.3.)

◆ Again, we *can use the units in the problem* to find our answer. How many tables are there? (4 tables.) How many chairs fit around these tables? (8 chairs.) How many people could sit around these tables? (8 people.) Is there enough room for the 12 people to sit at these tables? (No, only 8 people would fit.)

Continue with the other pages in the story in which more guests arrive and more tables are pushed together. Introduce your students to columns and

Figure 2.3

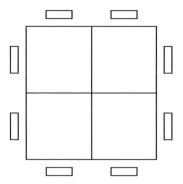

rows created from pushing the tables together to form various arrays. Create a chart (see Table 2.3) to keep track of each table arrangement, number of rows and columns in each array, number of tables and the number of people who could sit at each table when the square tables are pushed together.

If your students are ready to learn about area and perimeter you can introduce these concepts with the information in Table 2.3. The number of tables represents the area of each rectangle, which they can also calculate

Table 2.3

Table Arrangement	Number of Rows	Number of Columns	Number of Square Tables = Area	Number of Chairs (people) Around Tables = Perimeter
⬚⬚	1	2	2	6
⊞	2	2	4	8
⊞⊞	2	3	6	10
⊞⊞⊞	2	4	8	12
⊞ ⊞	2 2	2 2	8	16
⬚⬚⬚⬚⬚⬚⬚⬚	1	8	8	18
⬚⬚⬚⬚ ⬚⬚⬚⬚	1 1	4 4	8	20
▯ ▯ ▯ ▯	2 2 2 2	1 1 1 1	8	24

by multiplying the number of rows by the number of columns for each rectangle (in some examples there is more than one number for a row and column because there is more than one rectangle in the table arrangement). The number of chairs/people that fit around the table arrangement is the perimeter of the rectangle, which they can also calculate by adding up the number of chairs that would fit on each side of the rectangle. Students can use squares cut out of construction paper or use inch tiles and inch grid paper to create the arrays.

What Does This Standard Mean for Grades 3–5 Problem Solvers?

Students in intermediate grades who are successful at SMP 2 can make sense of the quantities in the problem and understand how they are related, such as in a multi-step problem that contains quantities representing various units. Students are able to step back and think about the context of the problem and, rather than relying on key words, can relate to the situation described in the problem to determine which operation(s) to employ. More advanced problem solvers can create equations with the unknown in a variety of positions in order to efficiently solve a problem. By fifth grade, students should be proficient with the properties of all four operations and be able to apply them to problem solving as needed.

In order to fully apply SMP 2 when approaching a word problem, students in intermediate grades should be able to apply the following *I Can* statements:

◆ *I can understand how the quantities in the problem are related.*
◆ *I can represent the context of the problem in order to create an equation.*
◆ *I can apply the properties of operations while solving the problem.*

Racing Around by Stuart J. Murphy

The characters in the book *Racing Around* (2002) want to enter a 15-kilometer bike race so they are riding their bikes around various locations and measuring their distance. This book shows places such as a rectangular athletic field and a zoo shaped as a pentagon, with distances marked off in kilometers for each side. Students can use the visuals and the measurements in the book to practice finding the perimeter of various plane figures and to explore the Associative Property as they create equations.

Measurement and Data 3.MD

Geometric measurement: recognize perimeter as an attribute of plane figures and distinguish between linear and area measure.
Solve real-world and mathematical problems involving perimeters of polygons, including finding the perimeter given the side lengths, finding an unknown side length and exhibiting rectangles with the same perimeter and different areas or with the same area and different perimeters.

As you read the book, encourage your students to calculate the distance each character rode by representing the perimeter of the various locations with an equation:

- Let's look at the page where Justin rides his bike around the athletic field, which is shaped like a rectangle. What do we know about the sides of a rectangle? (There are two long sides and two short sides; the opposite sides are congruent.)
- Justin is trying to show his siblings that he can ride 15 km in the bike race. How can we figure out how many kilometers he rode around the athletic field? (We can add up the measurements on each side of the field; we can find the perimeter of the field.)
- In order to find the distance around the field we would find the perimeter of the field. *We can represent the context of the problem in order to create an equation.* What information do we need to create an equation in order to calculate the perimeter of the field? (We need the length and the width of the rectangular field; we can use the measurements labeled on each side of the rectangular field.)
- Does the order of the addends matter in our equation if we are adding all four quantities? Why or why not? (It does not matter because we can add the numbers in any order and they will still have the same sum.) *We can apply properties of operations when we solve a problem.* What is the name of the property that represents this situation? (The Associative Property of Addition.)
- I want you to create at least two different equations for the perimeter of the athletic field, using addition to find your answer. ($1+1+2+2=6$; $1+2+1+2=6$; $2+2+1+1=6$; $2+1+1+2=6$; $2+1+2+1=6$.)
- *We can think about how the quantities in the problem are related.* Two sides are 1 km in distance and two of the sides are 2 km in distance. Can we create an equation for the perimeter of the field using

      multiplication in our equation? Why or why not? (Yes, because there are two sets of sides that are the same; we can multiply each length and width by 2.)

◆ Okay, so we know we can include multiplication because we can multiply the length by 2 and multiply the width by 2. Which multiplication facts would we use to represent the length and the width? (2×1 and 2×2.)

◆ We have to continue to *think about how the quantities in the problem are related*. What would we do with the products of these two multiplication facts? Do we need another operation in our equation? (We would have to add the products; we need to use addition in our equation.)

◆ This is an opportunity to use parentheses in our equation so we can clearly see that we should multiply the quantities then add those products to get our total distance. Our equation would be $(2 \times 1) + (2 \times 2) = 6$.

◆ Does the order of the factors matter in the parentheses? Why or why not? (It does not matter because we can multiply the numbers in any order and they will still have the same product.) *We can apply properties of operations when we solve a problem.* What is the name of the property that represents this situation? (The Associative Property of Multiplication.)

Continue reading the story and stop on page 12, which shows the distances Justin traveled on his bike as he rode around the zoo. It is shaped like a pentagon so students can attempt to create as many different addition equations as possible, applying the Associative Property of Addition, to represent the total number of kilometers around the zoo. Then challenge your students to create another equation with multiplication to represent the sides with the same distance. Remind them to use parentheses to separate the multiplication facts from the addition facts and they can apply the Associative Property of Multiplication to switch the order of the factors in their equation. The race course is marked on pages 28 and 29 with six different distances marked on the course. Your students can create equations independently to be used as a formative method of assessing their comprehension of perimeter and properties of operations.

Fractions, Decimals and Percents by David A. Adler

The setting of the book *Fractions, Decimals and Percents* (2010) is a county fair with numbers represented as fractions (as well as the equivalent decimal and percent on some pages). The narrator demonstrates fractions with denominators of 100 as they relate to prices of food, games and other situations common to a county fair. Students can see how prices can be represented in both decimals and fractions so they can begin to compare decimal fractions.

Number and Operations—Fractions 4.NF

Understand decimal notation for fractions, and compare decimal fractions.

Use decimal notation for fractions with denominators 10 or 100; compare two decimals to hundredths by reasoning about their size. Recognize that comparisons are valid only when the two decimals refer to the same whole. Record the results of comparisons with the symbols >, = or <, and justify the conclusions.

As you read the book, encourage your students to keep track of the equivalent fractions and decimals so they can begin to rewrite fractions in decimal notation:

◆ Let's look at the first page where the boy is buying cotton candy. How much does it cost? How do you know? (It costs 89¢. There is a sign on the counter with the price.)
◆ Can you find a fraction on this page that represents the same amount? (89/100) Yes, 89/100 is the same as 89¢. *We can think about how the quantities in the problem are related.* There are 100 cents in a dollar and the boy is paying only 89 cents so he is paying 89 out of 100 cents for the cotton candy.
◆ On the next page we learn we can write 89 cents as a decimal by putting a decimal point in front of the 89 so it will be 0.89, which we read as eighty-nine hundredths. That is also how we would read the fraction equivalent: 89/100 is eighty-nine hundredths.
◆ Now the boy wants to go to the Magic Show. How much does he have to pay and how do you know? (The sign says it is 41¢.) What is this amount written as a fraction? (41/100.) What is this amount written in decimal notation? (0.41.) How do we read both of those amounts? (Forty-one hundredths.)
◆ *We can represent the context of the problem to create an equation.* We know the fraction 41/100 is the same as, or equivalent to, the decimal notation 0.41 so we can write an equation: 41/100 = 0.41.

◆ On the next page there are two quarters and three pennies. How much money is this? (53¢.) How can we write that amount in both fraction and decimal notation keeping in mind that there are 100 cents in a dollar? (We can write the amount as 53/100 and 0.53.)

◆ There is another picture of coins—three nickels and two pennies. What is the amount of money these coins represent, what is the fraction of a dollar and what is the decimal notation? (It is 17¢ and it is 17/100 and it is 0.17.)

◆ Next, let's compare the decimals 0.53 and 0.17. We can compare them by reasoning about their size when they refer to the same whole. In this case they represent money and the whole is a dollar. Is one more than the other or are they the same and how do you know? (0.53 is more because 53¢ is more than 17¢.)

◆ Now *we can represent the context of the problem to create an equation.* We can use the greater and then less than symbols and the decimal notations to create an equation. We can write two equations: 0.53 > 0.17 and 0.17 < 0.53.

Continue reading the book and stop at pages where there are examples of decimals and fractions so your students can practice writing fractions in decimal notation. You can also have students find other opportunities to compare prices in decimal notation, write equations using the >, < and = symbols, and justify their answer.

Count to a Million by Jerry Pallotta

You can use the illustrations in the book *Count to a Million* (2003) as a visual way to show how quantities increase from ten to a million. On one page we can see 10 ladybugs, then on the next page we can see 100 ladybugs; we see 100 gumballs then we see 1,000 gumballs. On the pages are multiplication facts representing powers of 10 so students can see that a digit in one place represents 10 times as much as it represents in the place to its right.

Number and Operations in Base Ten 5.NBT

Understand the place value system.

Recognize that in a multi-digit number, a digit in one place represents 10 times as much as it represents in the place to its right and 1/10 of what it represents in the place to its left.

As you read the book, encourage your students to keep track of the multi-plication facts on each page so they can see how the quantity on one page is ten times as large as the quantity on the following page:

- ◆ Let's look at the page of 10 ladybugs. We see a multiplication fact on this page along with the illustration. It is $1 \times 10 = 10$. This equation represents the picture because the book started with the number 1, and there are 10 times as many ladybugs; this also represents one group of 10.
- ◆ On the next page we see 100 ladybugs and they are in groups of 10. What does the equation $10 \times 10 = 100$ represent on this page? (It represents 10 groups of 10 ladybugs; there are 10 times more ladybugs on this page than there were on the previous page.)
- ◆ On this page there are 100 gumballs and on the next page there are 1,000 gumballs. *We can understand how the quantities are related.* Why is the equation $100 \times 10 = 1,000$ on this page? (There are 10 groups of 100 gumballs; there are 10 times more gumballs on this page than there were on the previous page.)
- ◆ On this page there are 1,000 ants. *We can represent the context of the problem in order to create an equation.* If I told you there are 10,000 ants on the next page, what is the equation we would expect to see on that page and how do you know? (It would be $1,000 \times 10 = 10,000$ because there would be 10 times as many ants on that page; 10,000 is 10 times larger than 1,000; there would be 10 groups of 1,000 ants.)
- ◆ On this page there are 10,000 people running in a race. *We can represent the context of the problem in order to create an equation.* If I told you there are 100,000 people on the next page, what is the equation we would expect to see on that page and how do you know? (It would be $10,000 \times 10 = 100,000$ because there would be 10 times as many people on that page; 100,000 is 10 times larger than 10,000; there would be 10 groups of 10,000 people.)
- ◆ On this page there are 100,000 people in a stadium. *We can represent the context of the problem in order to create an equation.* If I told you there are 1,000,000 people on the next page, what is the equation we would expect to see on that page and how do you know? (It would be $100,000 \times 10 = 1,000,000$ because there would be 10 times as many people on that page; 1,000,000 is 10 times larger than 100,000; there would be 10 groups of 100,000 people.)

Your students can start with one million and write equations to produce numbers that are 1/10 of a quantity in order to understand that a digit in one place represents 1/10 of what it represents in the place to its left. They could use a calculator at first to see how they can either multiply by 1/10 or divide by 10 to find the answer, thus creating multiplication or division equations. Be sure to dispel the "rule" that students often use to multiply a number by 10 in which they simply add a 0 at the end of a number, as this does not always work, such as when multiplying numbers in decimal notation by 10.

Wrapping It Up

Students can begin moving to the abstract stage of problem solving once they have a handle on how to make sense of the quantities in the problem. They should be exposed to the various situations for each operation specified in Tables 1 and 2 in the Glossary of the Common Core State Standards for Mathematics (see Tables A1.1 and A1.2 in Appendix); only then can they compare the situation in the word problem to the known situations in order to decide which operation would be most efficient. SMP 2 also provides an opportunity for students to become exposed to properties of operations with the support of concrete objects or visuals so they can apply these properties in abstract form as they move through the grades into more advanced algebra.

3

Construct Arguments

SMP 3—Construct Viable Arguments and Critique the Reasoning of Others

How many times have you witnessed a student obtain the correct answer to a word problem because they just guessed which operation to use to solve the problem? Without even reading the words, students often search for the numbers and try to "do something" with them. As word problems become more complex—involving multiple numbers, units, steps and operations—children will find it more difficult to solve word problems by just guessing. If students can learn at an early age to justify their answer, as well as share their reason with others, they can build their repertoire of strategies and get in the habit of crafting an explanation.

What Does This Standard Mean for Grades K–2 Problem Solvers?

Students who are successful at SMP 3 construct an explanation to support their answer the minute they are deciding which operation to perform. They have a reason for their decision based on experience examining the various problem solving situations for addition and subtraction based on Table 1 in the Glossary of the Common Core State Standards for Mathematics (see Table A1.1 in Appendix). They can justify their answer with a drawing, a diagram, a concrete model or by acting it out. They know how to listen and ask questions when being exposed to the reasoning of their peers.

In order to fully apply SMP 3 when approaching a word problem, students should be able to take ownership of their procedures by using the following *I Can* statements:

◆ *I can explain my reason for my answer.*
◆ *I can use objects, drawings, tables and actions to represent the problem.*
◆ *I can listen and respond to the way others solved the same problem.*

How Many Mice? by Michael Garland

The book *How Many Mice?* (2007) can be used by students who are ready to craft an explanation for their reasoning and listen to the explanations of their peers. The story follows ten hungry mice that leave their home to gather their meal, encountering an abundance of food as well as several predators on their journey. You can encourage students to use the illustrations to count the mice and their pieces of food and answer the question posed about the mice or the food on each page. Your students can use the events in the story to represent addition and subtraction problems by looking at the illustrations, acting out situations, creating their own drawings and, when the students are ready, using equations to represent the word problems.

Operations and Algebraic Thinking K.OA

Understand addition as putting together and adding to, and understand subtraction as taking apart and taking from.

Represent addition and subtraction with objects, fingers, mental images, drawings, sounds, acting out situations, verbal explanations, expressions or equations.

As you read the book, encourage your students to explain how many mice and food are on each page by representing the problem on each page in various ways:

◆ Let's look at the page where the mice first find some food. Each mouse has a cherry. The question on this page is, "How many cherries can you count?" (10.) We found our answer by counting the cherries on these pages and we counted from 1 to 10. As problem solvers, we *can explain the reason for our answer.*
◆ Sometimes we *can use objects to represent a problem.* I have 10 counters to represent the 10 cherries that we counted on the first page. But now we see the crows stole 4 of the cherries. The question on this page is, "How many cherries do the mice have now?" What can we do to find our answer? (We can take away 4 of the cherries and count how many are left.)

◆ Yes, we should take 4 of the cherries from our group of 10 cherries. What operation is being modeled here? How do you know? (There were 10 cherries and the crows stole 4 of the cherries. That would be subtraction because cherries are taken from the group.)

◆ Now we are ready to represent the problem with symbols. I will use the numbers 10 and 4 and the subtraction and equal symbol. I will also use the letter n to represent the number that we will get for our answer. Here is the equation, $10-4=n$. What is the answer to this equation and how do you know? (The answer is 6 because after we take away 4 we can count the rest and there are 6 cherries left.)

◆ Sometimes we *can use drawings to represent a problem.* I am going to draw 6 cherries because we said the mice only have 6 cherries left after the crows stole 4 of their cherries. But now we see some of the mice found a tomato plant. The story says they picked 2 tomatoes so I am going to draw 2 tomatoes next to the 6 cherries. The question on the page is, "How many pieces of food do the mice have?" What can we do to find our answer? (We can count the cherries and tomatoes.)

◆ Yes, we should count all of the cherries and the tomatoes in my drawing. What operation is being modeled here? How do you know? (It is addition; we are adding tomatoes to the group of cherries and counting all of them.)

◆ Now we are ready to represent the problem with symbols. I will use the numbers 6 and 2 and the addition and equal symbol. I will also use the letter n to represent the number that we will get for our answer. Here is the equation, $6+2=n$. What is the answer to this equation and how do you know? (The answer is 8 because after we count 6 cherries we keep counting the tomatoes and we counted to 8.)

◆ Good problem solvers *can listen to the way others solved the same problem.* On the next page we see the mice with the 6 cherries, the 2 tomatoes and some corn. I am going to give you and a partner some counters so you can find the answer to the problem on this page, "Can you count all of the pieces of food?" First count out 6 counters, then count out 2 counters, then count out 4 counters for the corn. Tell your partner how you will find your answer and then see if you and your partner have the same answer and if you found your answer the same way.

As you continue reading the story, your students can keep track of the changing number of food items held by the mice prompted by the questions on each page. They could draw the objects, act out the situation or use counters and should also describe the situations to help them decide if addition or subtraction would be the most efficient operation to solve the problem. Then they can create an equation to represent the problem with symbols. Have students practice explaining to the class how they arrived at their answer.

Mall Mania by Stuart J. Murphy

The book *Mall Mania* (2006) is about a group of five children on a chess club who are counting the number of people entering a mall in order to give a prize to the 100th shopper. Four of the children are stationed at different mall entrances while the team captain is inside with a walkie-talkie, keeping track of the totals reported each hour. As the children call in their numbers, they use different strategies to add the numbers. Your students can practice using and sharing various strategies for addition.

Number and Operations in Base Ten 1.NBT

Use place value understanding and properties of operations to add and subtract.

Add within 100, including adding a two-digit number and a one-digit number, and adding a two-digit number and a multiple of 10, using concrete models, or drawings and strategies based on place value, properties of operations and/or the relationship between addition and subtraction; relate the strategy to a written method and explain the reasoning used.

As you read the book, stop at various pages in order to allow your students to engage in the following activity, focusing on the *I Can* statements:

- ◆ On page 10, the children counted the following number of shoppers at the four mall entrances: 7, 4, 3 and 2. Good problem solvers *can explain their reason for their answer.* Add up the numbers and be ready to tell the class how you added the numbers together.
- ◆ On page 12, we can see that Nicole and Gabby both have 16 as their answer but they used different strategies to add them up. Let's see if you *can listen and respond to the way others solved the same problem.* Listen while I read their explanations. Respond if you used a similar strategy to add the numbers. Did anyone have a different strategy?
- ◆ Who *can use objects, drawings, diagrams and actions to represent the problem?* Please come up to the board and show us how you added the numbers.

◆ On page 14, the children counted 8, 8, 7 and 7 shoppers. Add up these numbers and see if you *can explain your reason for your answer*.

◆ On page 16, we can see that Jonathan and Steven both have 30 as their answer but they used different strategies to add them up. Let's see if you *can listen and respond to the way others solved the same problem*. Listen while I read their explanations. Respond if you used a similar strategy to add the numbers. Did anyone have a different strategy?

◆ On page 18, the children counted 8, 9, 7 and 8 shoppers. Add up these numbers and see if you *can explain your reason for your answer* by writing a sentence or two explaining how you added the numbers together.

◆ On page 20, we can see that Steven and Gabby both have 32 as their answer but they used different strategies to add them up. Let's see if you *can listen and respond to the way others solved the same problem*. Listen while I read their explanations. Respond if you used a similar strategy to add the numbers. Did anyone have a different strategy?

◆ On page 23, the children counted 5, 5, 5 and 6 shoppers. Add up these numbers and see if you *can explain your reason for your answer* by using the chart paper for your group to write your number sentence and your explanation of how you added the numbers together.

◆ On page 24, we can see that Jonathan and Nicole both have 21 as their answer but they used different strategies to add them up. Let's see if you *can listen and respond to the way others solved the same problem*. Let's look at the chart paper for each group while someone at the group reads the sentences. Respond if someone at your group used a similar strategy to add the numbers. Did anyone in your group use a different strategy?

Summarize the activity by having students name some of the strategies used in the book to add the numbers. How many different strategies were used? Who used a strategy that was used in the book? Were there strategies used by your students that were not used in the book? How did the explanation and the illustrations of the strategies help everyone understand them?

How Many Seeds in a Pumpkin? by Margaret McNamara

In the book *How Many Seeds in a Pumpkin?* (2007), Mr. Tiffin asks his first grade class to estimate the number of seeds in each of the three pumpkins sitting in front of the class—one small, one medium and one big pumpkin. The students in the story provide reasons for their estimate, then count the seeds in each pumpkin by twos, by fives and by tens. Your students can practice skip-counting and comparing three-digit numbers with the seeds in the story.

Number and Operations in Base Ten 2.NBT

Understand place value.

Count within 1,000; skip-count by fives, tens and hundreds.

Compare two three-digit numbers based on meanings of the hundreds, tens, and ones digits, using >, = and < symbols to record the results of comparisons.

As you read the book, engage students in a discussion by asking the following questions based on the *I Can* statements:

◆ In this story, Charlie and his classmates are comparing the size of the three pumpkins that their teacher, Mr. Tiffin, put on his desk one morning. Mr. Tiffin wants them to estimate the number of seeds in each pumpkin. As we become better problem solvers we *can explain the reason for our answer*. First let's see how the characters in the story explain the reasons for their answers.

◆ How many seeds did Robert, Elinor and Anna say were in the pumpkin and why did they give that number? (Robert said there were one million seeds in the biggest one because the biggest one has the most. Elinor said the medium one has 500 because she always sounded like she knew what she was talking about. Anna said the tiny one has 22 because she liked even numbers.)

◆ When the children were opening up the pumpkins, what did Robert say as they all looked inside the pumpkins? (He said the big one definitely has the most.) Why do you think he says this? (He was looking at the seeds inside so there must have been a lot in the big one.) How could the students *use objects, drawings, tables or actions to represent the problem?* (They can take the seeds out and count them.) How do they decide to count the seeds? (They lay out all of the seeds on a table. Some students count them by twos, by fives and by tens.)

◆ Let's look at the page where one of the groups, the Twos Club, laid out their seeds. How did they arrange their seeds so they could count them easily? (They put them in pairs so there are 2 seeds together.) Let's count the first row of seeds together by twos, 2, 4 … 34. It says there are 170 pairs of seeds so we'll find out later in the story how many seeds there are in the biggest pumpkin.

◆ Let's look at the page where one of the groups, the Fives Club, laid out their seeds. How did they arrange their seeds so they could count them easily? (They put them in groups of 5 seeds together.) Let's count the first row of seeds together by fives, 5, 10 … 45. It says there are 63 groups of 5 seeds and 1 seed left over so we'll find out later in the story how many seeds there are in the medium pumpkin.

◆ Let's look at the page where Charlie, the Tens Club, laid out his seeds. How did he arrange the seeds so he could count them easily? (He put them in 2 rows of 5 so there are 10 seeds together.) Let's count the first row of seeds together by tens, 10, 20 … 70. It says there are 35 groups of 10 seeds so we'll find out later in the story how many seeds there are in the smallest pumpkin.

◆ It took the Twos Club a long time to count all of the seeds by twos but they ended up counting 340 seeds in the biggest pumpkin. We *can use a table to represent the problem.* Let's keep track of how many hundreds, tens and ones are in each pumpkin. (See Table 3.1 for the biggest pumpkin.)

◆ It did not take as long for the Fives Club to count all of the seeds by fives, which was 315, and then they added the 1 seed left over to count a total of 316 seeds in the medium pumpkin. We *can use a table to represent the problem.* Let's add the hundreds, tens and ones for the medium pumpkin. (See Table 3.2.)

Table 3.1

Size of Pumpkin	Hundreds	Tens	Ones
Big	3	4	0
Medium			
Small			

Table 3.2

Size of Pumpkin	Hundreds	Tens	Ones
Big	3	4	0
Medium	3	1	6
Small			

◆ Now we're going to compare the number of seeds in the big pumpkin and in the medium pumpkin. We can use our table to compare the hundreds, tens and ones in each number. In order to decide which number is larger, we have to first look at the hundreds. How many hundreds are in the big pumpkin? (3) How many hundreds are in the medium pumpkin? (3) Are there more hundreds in the big pumpkin or in the medium pumpkin and how do you know? (They have the same because if we look at the table, there are 3 hundreds in each pumpkin.)

◆ So we have to look at the tens next. How many tens are in the big pumpkin? (4) How many tens are in the medium pumpkin? (1) So are there more tens in the big pumpkin or in the medium pumpkin and how do you know? (There are more tens in the big pumpkin because 4 tens is more than 1 ten.) We don't have to look at the ones now because we have enough information to decide that 340 is bigger than 316. We can use symbols to record the results: 340 > 316.

◆ But Charlie still has to count his seeds. It did not take long at all for him to count 35 groups of 10, giving him a total of 350 seeds in the smallest pumpkin. We *can use a table to represent the problem.* Let's add the hundreds, tens and ones for the smallest pumpkin. (See Table 3.3.)

◆ Now we're going to compare the number of seeds in the big pumpkin and in the smallest pumpkin. We can use our table to compare the hundreds, tens and ones in each number. In order to decide which number is larger, we have to first look at the hundreds. How many hundreds are in the big pumpkin? (3) How many hundreds are in the smallest pumpkin? (3) Are there more hundreds in the big pumpkin or in the small pumpkin and how do you know? (They have the same because if we look at the table, there are 3 hundreds in each pumpkin.)

◆ So we have to look at the tens next. How many tens are in the big pumpkin? (4) How many tens are in the small pumpkin? (5) So are

Table 3.3

Size of Pumpkin	Hundreds	Tens	Ones
Big	3	4	0
Medium	3	1	6
Small	3	5	0

there more tens in the big pumpkin or in the small pumpkin and how do you know? (There are more tens in the small pumpkin because 5 tens is more than 4 tens.) We don't have to look at the ones now because we have enough information to decide that 350 is bigger than 340. We can use symbols to record the results: 350 > 340.

◆ What does Mr. Tiffin say about being able to tell how many seeds are in a pumpkin? Think about how you *can listen and respond to the way others solved the same problem.* (The small pumpkin had the most seeds. Mr. Tiffin said you can't tell until you open it.)

◆ Mr. Tiffin said there are some clues. What were they? Again, think about how you *can listen and respond to the way others solved the same problem.* (For each line on the outside there is a row of seeds on the inside. The longer the pumpkin grows the more lines it gets and its skin gets darker. That's why the smallest pumpkin had the most seeds. It was dark orange and had the most lines on the outside.)

After reading the book, have your students practice comparing other three-digit numbers based on real-world examples, such as the number of students in specific grade levels in your school or from articles in the newspaper. They can make a table to record the number of hundreds, tens and one in each three-digit number or create a model with base ten blocks. Then have them record their results with the greater than symbol. They can also practice justifying their reason for choosing the larger number based on the hundreds, tens and ones in each number.

What Does This Standard Mean for Grades 3–5 Problem Solvers?

Students in intermediate grades who are successful at SMP 3 can justify their problem solving strategies and their answer using accurate mathematical definitions as well as draw upon their prior experience with similar problems. They can apply their strategy to the particular operation with the support of the various problem solving situations for multiplication and division based on Table 2 in the Glossary of the Common Core State Standards for Mathematics (see Table A1.2 in Appendix). The developing and advanced problem solvers can analyze the justification of their peers and compare it to their own.

In order to fully apply SMP 3 when approaching a word problem, children in intermediate grades should be able to apply the following *I Can* statements:

◆ *I can use assumptions, definitions and prior results to construct my argument.*

◆ *I can analyze the arguments of others in response to the way they solved the same problem.*

◆ *I can create counterexamples and identify flaws in an argument.*

Full House by D.A. Dodds

The book *Full House* (2007) is set at The Strawberry Inn where Miss Bloom puts each of her five guests in a cozy room. As each room is filled, the fraction (out of 6) is shown at the bottom of the page with a poem describing the numerator and denominator. Your students will be able to see how the denominator of six does not change but the numerator increases as each room, including Miss Bloom's, is filled by the end of the night.

Number and Operations—Fractions 3.NF

Develop understanding of fractions as numbers.

Understand a fraction $1/b$ as the quantity formed by 1 part when a whole is partitioned into b equal parts; understand a fraction a/b as the quantity formed by a parts of size $1/b$.

As you read the book, stop at various pages in order to allow your students to engage in the following activity, focusing on the *I Can* statements:

◆ On the first page, we are given the information that there are six rooms at the Strawberry Inn, one for each of the five guests and one room for Miss Bloom. Then on the next page we see Miss Bloom escorting her first guest, Captain Duffy, to his room. There is a fraction at the bottom of the page with a poem that gives us the explanation of the numerator and the denominator. Why is the fraction 1/6? (There are six rooms at the Inn and there is one room filled.) Yes, we can think of the six rooms as the whole because those are all of the rooms at the Inn. We can think of the one room as the part because it is only part of the whole set of rooms at the Inn.

◆ Before we turn the page to see the next guest be escorted to her room, I want you to think about what fraction might appear on this page and why you think so. As good problem solvers we *can use assumptions, definitions and prior results to construct an argument.* Be sure to explain how you came up with your number for the numerator and denominator. (Students should share their fraction and justify why they think their answer is correct.)

◆ You were correct if you stated the fraction on the next page is 2/6. Some of you explained the numerator is a two because now two rooms are filled and the denominator remains a six because there are six rooms at the Inn. You used definitions of numerator and denominator or you based your argument on the results from the prior example. Some of you had an incorrect answer or you had a correct answer but could not fully explain your answer.

◆ Now you will try it again with the next page when Miss Bloom escorts her third guest to his room. Be sure to *use assumptions, definitions and prior results to construct your argument.* This time you will write down your argument and compare it with others in your group so you *can analyze the arguments of others in response to the way they solved the same problem.* Be prepared to share with the class how all of your arguments were similar and how they were different.

◆ You were correct if you stated the fraction on the next page is 3/6. I heard your arguments stating the numerator is a three because now three rooms are filled and the denominator remains a six because there are six rooms at the Inn. You used definitions of numerator and denominator or you based your argument on the results from the prior example. You also compared and contrasted your argument with that of your group members. Now we want to demonstrate how we *can create counterexamples and identify flaws in an argument.*

◆ I am going to give you an argument for the problem we just completed from a student who is not in our class and I want you to find a flaw in his argument:

I think on the next page the fraction will be 3/6 because there are three empty rooms so that is why the numerator is three. There are six rooms in all so that's why the denominator is six.

You were correct if you stated the flaw is with his explanation of why the numerator is a three. In the story the numerator changes as the rooms are filled, not as they are empty. But his explanation for why the denominator remains a six is correct—it's because there are six rooms in all at the Inn. This flawed argument is a common one because the answer of 3/6 is correct but due to his argument, this student does not fully understand how to determine the numerator.

Continue to have students prepare their arguments for the remaining fractions in the story based on the number of rooms filled by guests. At the end of the story, each guest and Miss Bloom eat one of the six pieces of cake in the kitchen. You can create another fictional flawed argument or use the scenario as an assessment to determine the level of understanding of your students as they write an argument for the change in the numerator as each piece of cake is eaten.

The Warlord's Puzzle by Virginia Walton Pilegard

The Warlord's Puzzle (2000) is a tale set in China about a warlord who loved beautiful and rare objects. An artist presents the warlord with a unique blue tile but drops it while showing it off. The tile breaks into seven pieces—a parallelogram, a square and five triangles. Your students may recognize the seven pieces as the set of tangram shapes which can be used to explore properties of two-dimensional shapes.

Geometry 4.G

Draw and identify lines and angles, and classify shapes by properties of their lines and angles.

Classify two-dimensional figures based on the presence or absence of parallel or perpendicular lines, or the presence or absence of angles of a specified size. Recognize right triangles as a category, and identify right triangles.

As you read the book, provide your students a set of tangrams to explore while focusing on the *I Can* statements:

◆ You will each have a set of tangram pieces to use as we read the story. In order to be proficient at constructing viable arguments we *can use assumptions, definitions and prior results to construct our argument.* I want you and your partner to make a list of all of the properties of a parallelogram, square and triangle focusing on the sides and angles. Then discuss the properties of these specific shapes as they are represented in the tangram set and we will create a chart with the information. (See Table 3.4.)

◆ Now we can use the properties in our chart as we construct, critique and defend arguments about these two-dimensional shapes. When the artist drops the tile it breaks into seven pieces: a parallelogram, a square and five triangles. What is the problem that the warlord wants solved? (He wants someone to put the seven pieces back together into the shape of the square tile.)

Table 3.4

Shape	Properties of Sides	Properties of Angles	Shape in Tangram Set
Parallelogram	Opposite sides are congruent. Opposite sides are parallel. Called a quadrilateral.	Opposite angles are congruent. Consecutive angles are supplementary.	Two acute and two obtuse angles. Two shorter sides and two longer sides. Longer side is same length of diagonal of small triangle.
Square	Opposite sides are congruent. Opposite sides are parallel. Called a quadrilateral. All sides are congruent.	Opposite angles are congruent. Consecutive angles are supplementary. All angles are right angles (90°).	Same size as two of the smaller triangles put together at diagonals. Side is same length as leg of small triangle.
Triangles	There are three sides.	The sum of all angles is 180°.	Two large right triangles. Two small right triangles. One medium right triangle.

◆ We see as we read the story there are many people lined up to try to solve the problem. Finally a peasant and his young son join the line and when their turn finally came what did the boy think about as he played with the seven pieces? (The pieces reminded him of objects; the medium triangle was like his hat, the two large triangles put together formed his father's hat, the little triangles were fish, the parallelogram was the weave in their fishing net and the square was a box for the fish.)

◆ On the last page we see the seven pieces put together into the original square tile. I will close the book and give you one minute to see if you can put the seven pieces together just like the boy did in the story. Then I will give you a sheet with a picture of the tangram pieces put into the tile so you can check your results. (See Figure 3.1.) Then I will give you some problems to solve using the picture and the table of properties we created.

◆ The warlord in the story is happy the tile has been restored but he wants to learn more about the design created by the seven smaller pieces. We are going to write a letter to the warlord explaining the design is actually made up of seven shapes. We will create the letter together as a class using language that is precise and easy to understand.

Figure 3.1

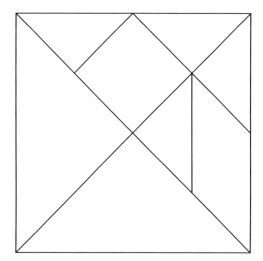

◆ Your task is to create a draft letter providing the name and explaining the properties of each shape. You have to use at least these six terms in your letter: parallel, perpendicular, sides, angles, opposite and right angle. You will include the sheet with the picture of the tangram pieces in your draft letter so you can label each shape and its properties. When you are done you will share your draft letter and labeled picture with a partner and defend why your letter should be the one we send to the warlord.

◆ As you listen to your partner's argument, critique their argument by writing down at least one reason why it's a strong argument and one reason why it is not. This will be your opportunity to demonstrate that you *can analyze the argument of others in response to the way they solved the same problem.*

◆ Once each set of partners has shared their own argument and their critique of their partner's argument, use those reasons to strengthen your own argument, which you will share with a new partner.

Your students should continuously improve their definitions by using precise and accurate language so their letter could easily be understood by an outside reader (which in this case is the warlord). They should also strengthen their argument by using the critique of their partners to improve their argument. You can also challenge your students by having them label each angle in the picture of the tile (Figure 3.1) using what they know about right angles, supplementary angles and perpendicular lines.

The Hershey's Milk Chocolate Fractions Book by Jerry Pallotta

Children will be eager to see how they can use a Hershey chocolate bar to learn about fractions in the book *The Hershey's Milk Chocolate Fractions Book* (1999). On each page the Hershey bar is used as the whole and the 12 equal sections of the candy bar represent the fractional pieces. Your students can explore the concept of equivalent fractions using the visual representation of the Hershey chocolate bar so they can begin to add and subtract fractions.

Number and Operations—Fractions 5.NF

Use equivalent fractions as a strategy to add and subtract fractions.
Add and subtract fractions with unlike denominators by replacing given fractions with equivalent fractions. Solve word problems involving addition and subtraction of fractions referring to the same whole, including cases of unlike denominators.

As you read the book, stop at various pages in order to allow your students to use their own set of 12 fraction pieces (made out of brown construction paper), representing the 12 sections of a Hershey's chocolate bar, as they use the *I Can* statements:

- ◆ Let's put our 12 fraction pieces into the shape of a Hershey's chocolate bar. We can see in the book there are 3 rows with 4 pieces in each row. This chocolate bar represents the whole, which we can rewrite as 12/12 because there are 12 equal pieces and we have all 12 pieces. We have learned in the past that when the numerator and denominator are the same, the fraction is equivalent to 1 whole. So, 12/12=1.
- ◆ Now let's remove 1 of the 12 pieces. How many pieces did we remove, stated as a fractional representation? (We removed 1/12 of the candy bar.) How many pieces do we have left? (We have 11/12 left.) We can make an addition problem with this information and it should equal 1. Let's try it: 1/12+11/12=1. Is this true? How do we know? Remember, good problem solvers *can use assumptions, definitions and prior results to construct an argument.* (Yes, it's true because when we have a common denominator we just add the numerators; so we add 1+11 and we get 12; we know that 12/12=1.)
- ◆ We are going to put that 1 piece back into our candy bar so we have all 12 pieces again. Now take your pieces and separate them so half of your pieces are on one side of your desk and the other half of

your pieces are on the other side of your desk. Let's just look at half of our candy bar. How many pieces do we have? Can we state our answer as a fractional representation in more than one way? (We have 1/2 of the candy bar; we have 6/12 of the candy bar.)

◆ As stated earlier, we can make an addition problem with this information and it should equal 1. Let's try it: $1/2 + 6/12 = 1$. Can you create an argument proving this is true? Remember, good problem solvers *can use assumptions, definitions and prior results to construct an argument.* I want you to write down your argument and share it with someone sitting next to you. We can also *analyze the argument of others in response to the way they solved the same problem* so let's practice doing that when we share our arguments with a peer. You can discuss how your arguments were similar and how they were different but also look for the following: precise language, definitions and a thorough explanation with visuals, such as an equation.

◆ Many of you included definitions of unlike denominators and explained how to create an equivalent fraction for 1/2 so you ended up adding $6/12 + 6/12$. You shared your arguments with a peer and compared and contrasted your arguments during your analysis. We know how to *create counterexamples and identify flaws in an argument* so during your analysis you should have critiqued each other's argument and pointed out any flaws.

◆ I am going to give you another equation to solve using your fractional pieces. You will also have to construct an argument and critique at least one other person's argument. Hopefully you heard some thorough arguments that you can use as a model, or you found flaws in your peer's argument, or in your own, from which you can learn.

◆ Here is the problem: $1/3 + 8/12 = ?$ You will have to use your fractional pieces to represent each fraction, to make an equivalent fraction, and in your argument.

There are other examples in the book which include adding fractions with unlike denominators that your students could use to practice using visual fraction models as well as equations to represent a problem. Students can also create equivalent fractions for the examples on each page, using their 12 fractional pieces. Finally, have students use the pieces to create subtraction problems with like and unlike denominators.

Wrapping It Up

Students in early grades are learning to use objects, drawings, diagrams and actions to represent word problems while they are building up their speaking, reading, writing and listening skills. They can practice their speaking skills by explaining their reason for their answers. As their literacy skills increase in the intermediate grades, students can label their drawings and diagrams and write sentences to explain their reason. Students can practice explaining their representation to their peers and when they are able to respond critically to each other's explanation, they are demonstrating they are able to analyze arguments.

4

Create a Model

SMP 4—Model with Mathematics

Most teachers will agree students should engage in problem solving related to their lives, yet many teachers only use word problems as they are written in their math curriculum. It is rare to find that every word problem is relevant to all students in your classroom, so care must be taken to individualize or modify word problems at times in order to fit the lives of your students. This will make it easier for your students to have some background knowledge related to the word problem as well as understand the language and context of the problem in order to work through the problem solving process.

What Does This Standard Mean for Grades K–2 Problem Solvers?

Students who are successful at SMP 4 in early grades are starting to see how word problems can be connected to their daily lives. They can find the quantities they need within a word problem and are becoming proficient in their knowledge of the addition and subtraction situations so they can represent those quantities. They can simplify a problem in order to make sense of the quantities and their relationships. They can also look back at the original problem to determine if the answer makes sense.

In order to fully apply SMP 4 when approaching a word problem, students should be able to take ownership of their procedures by using the following *I Can* statements:

◆ *I can solve problems in everyday life.*
◆ *I can identify important quantities and represent their relationships.*

◆ *I can simplify a problem.*
◆ *I can reflect on the results to see if they make sense.*

The Doorbell Rang by Pat Hutchins

In the book *The Doorbell Rang* (1986), a mother is baking chocolate chip cookies and offers a plate of the cookies to her two children, Victoria and Sam. Just as the children decide to share the 12 cookies among themselves, the doorbell rings and two more children come in. The story continues as more children arrive and Sam and Victoria agree to share the 12 cookies evenly among the group of children. At the end, Grandma rings the doorbell with another batch of cookies. You can use the examples of sharing the cookies to help your students decompose into smaller shares.

Operations and Algebraic Thinking K.OA

Understand addition as putting together and adding to, and understand subtraction as taking apart and taking from.
Represent addition and subtraction with objects, fingers, mental images, drawings, sounds, acting out situations, verbal explanations, expressions or equations.

As you read the book, engage students in the following activity based on some of the *I Can* statements:

◆ "How many of you have eaten cookies or some other type of treat that you had to share with others? It is good manners to offer treats to others rather than to eat in front of someone who does not have the treat. It is also good manners to share the treats so each person has an equal amount. In this book, we'll see two children, Victoria and Sam, who show they have very good manners by sharing the plate of cookies with their friends. This is one way we *can solve problems in everyday life.*"
◆ After reading the first two pages, stop and ask, "How do we know the number of cookies that are on the plate? Sam and Victoria say they will get 6 cookies each." Have the students turn to a partner and discuss how they can figure out the total number of cookies on the plate while you pass out a baggie with 12 brown circles representing the cookies. Allow students to tell you how many cookies they think are on the plate, accepting all answers and writing them on the board. Ask students to justify their answers.

◆ "I know there are 12 cookies because the 2 children each get 6 cookies. *I have identified these important quantities so now we can represent their relationships* with the equation, $6+6=12$. We know that Sam will get 6 cookies, so this is what the first 6 represents in our equation. We also know that Victoria will get 6 cookies, so this is what the other 6 represents in our equation. We know that 6 plus 6 equals 12. So there are 12 cookies on the plate." Draw a picture to show how the 12 cookies are shared among the two children (see Figure 4.1).

◆ Allow the students to use the cookie counters and make 2 piles of 6 cookies. Then tell them to count all of the cookies *to see if the results make sense.* They should be able to justify that 6 plus 6 equals 12. Continue reading the story and stop after the next set of pages when Tom and Hannah come to the door and Ma says they have to share the cookies.

◆ "There were only 2 children sharing the 12 cookies at the beginning of the story but now there are 2 more children at the door. How many children are there now?" (4) "Use your cookie counters to figure out how many cookies each child will get if they share the 12 cookies equally. You can use a dry erase board to draw 4 plates in order to evenly divide up the 12 cookies."

◆ Have a student draw the representation on the board, document camera or Smart Board to demonstrate how to divide up the 12 cookies among the 4 children (see Figure 4.2). Have the other

Figure 4.1

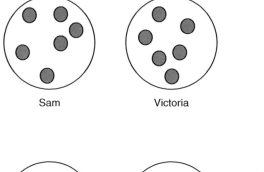

Sam   Victoria

Figure 4.2

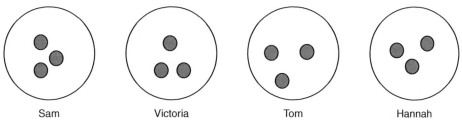

Sam   Victoria   Tom   Hannah

students check their physical model to see if they are correct. "Did Sam and Victoria get the same amount of cookies, more cookies or fewer cookies when they had to share them with Tom and Hannah?" (They get fewer cookies.)

On the next page, your students can check their answer as Sam and Victoria say they will each get 3 cookies. "Let's see if we *can reflect on the results to see if they make sense.* In this part of the story, the numbers of cookies are the *important quantities and that is what we can represent* with our equation, $3+3+3+3=12$."

- The doorbell rings again and 2 more children come in. "There were 4 children and now 2 more children have come to the door but there are still 12 cookies. In this part of the story, the numbers of children are the *important quantities and that is what we can represent* with an equation. Show me an equation to represent the number of children who now have to share the cookies." The students can use their dry erase board to write the equation $4+2=?$ Call on a student to explain what the 4 represents (the number of children already in the kitchen) and what the 2 represents (the number of children at the door). "We can count the children in the picture or solve it by adding 4 plus 2. There are 6 children now. Use your cookie counters to figure out how many cookies each child will get if they share the 12 cookies equally. You can use the dry erase board to draw 6 plates to evenly divide up the 12 cookies. Do you think Sam and Victoria will get the same amount of cookies, more cookies or fewer cookies now?" (They will get fewer cookies.) "How do you know?" (Every time they share cookies with more people, everyone gets fewer cookies.)
- Have a student draw the representation on the board, document camera or Smart Board to demonstrate how to divide up the 12 cookies among the 6 children (see Figure 4.3). Have the other students check their physical model to see if they are correct.
- "Let's see if we *can reflect on the results to see if they make sense.* In this part of the story, the numbers of cookies are the *important quantities and that is what we can represent* with our equation, $2+2+2+2+2+2=12$. As I read the next page, we can check our answer. Yes, we are correct because Sam and Victoria say they will each get 2 cookies."
- The doorbell rings again and 6 more children come in. "There were 6 children and now 6 more children have come to the door but there are still 12 cookies. In this part of the story, the numbers of

Figure 4.3

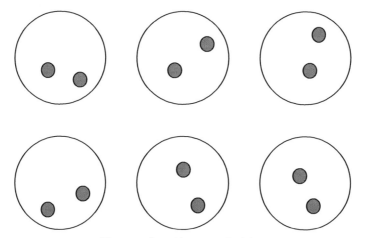

There are 2 cookies on each plate.

children are the *important quantities and that is what we can represent* with an equation. Show me an equation to represent the number of children who now have to share the cookies." The students can use their dry erase board to write the equation, $6+6=?$ Call on a student to explain what the first 6 represents (the number of children already in the kitchen) and what the other 6 represents (the number of children at the door). "We can count the children in the picture or solve it by adding 6 plus 6. There are 12 children now. Use your cookie counters to figure out how many cookies each child will get if they share the 12 cookies equally. If you need to use the dry erase board you will have to draw 12 plates to evenly divide up the 12 cookies."

◆ Have a different student draw the representation on the board, document camera or Smart Board to demonstrate how to divide up the 12 cookies among the 12 children (see Figure 4.4). Have the other students check their physical model to see if they are correct.

◆ "Let's see if we *can reflect on the results to see if they make sense*. In this part of the story, the numbers of cookies are the *important quantities and that is what we can represent* with our model of the 12 plates and 1 cookie on each plate. As I read the next page, we can check our answer. Yes, we are correct because Sam and Victoria say they will each get 1 cookie. What would happen if more children come to the door?" (They can break the cookies in half to share them. They can hurry and eat the cookies before Ma opens the door. They can ask the children to come back later when Ma makes more cookies.)

Figure 4.4

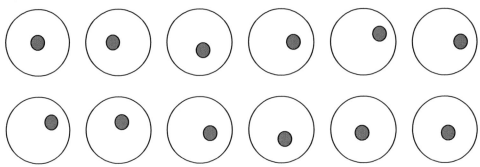

There is 1 cookie on each plate.

The doorbell does ring again but it is Grandma with another tray of cookies. Now students can practice other combinations of cookies to be shared equally among the children. Your students can relate to the real life problem in this story even if their family does not bake chocolate chip cookies by discussing other treats their family cooks, bakes or purchases. You can create word problems related to the story and allow students to use a physical model with counters, draw a visual model or work only with an equation, reinforcing the concept that as the number of people increases, each person's share gets smaller.

Bigger, Better, Best! by Stuart J. Murphy

Another familiar context to use with students besides the kitchen is their bedroom. In the book *Bigger, Better, Best!* (2002), three young siblings are constantly arguing about whose belongings are best. One day their parents announce they will be moving to a new house and will each have their own bedroom. When they go to the new house to select their bedrooms, they argue over which bedroom window is the biggest. Their mother suggests covering each window with sheets of paper to see which window has the largest area. When they argue about which bedroom is the biggest, their father suggests covering the floor with newspaper to see which room has the largest area. Your students can explore the concept of using rectangles to compose new shapes as well as become introduced to the concept of area.

Geometry 1.G

Reason with shapes and their attributes.

Compose two-dimensional shapes or three-dimensional shapes to create a composite shape, and compose new shapes from the composite shape.

As you read the book, engage students in the following activity based on the *I Can* statements:

◆ "How many of you have your own bedroom? Who has to share a bedroom? In this story, Jill and Jenny are sisters who share a bedroom. Sometimes they argue with their brother, Jeff, to compare their belongings, like a backpack or book, to see whose are better. Maybe you argue with a brother, sister, cousin, neighbor or friend too. Let's see how their parents help them settle their arguments when they go see their new house."

◆ Read the story and stop on page 13 when Jeff and Jenny begin arguing over whose bedroom window is bigger. "What was their mother's suggestion for checking to see which window is bigger?" (She told them to use paper to cover their windows. The window that uses more paper to cover it is the one that has the larger area.) Introduce the term *area*, defined as the number of square units to cover the inside of a shape.

◆ "Yes, by using the same size sheets of paper, they can see if it takes more sheets to cover one of the windows. This is one way we *can solve a problem in everyday life.*"

◆ Read through page 15 where Jeff completely covers the inside of his window with the sheets of paper. "How many sheets of paper did Jeff use to cover his window? Let's see if we can *identify important quantities* on this page *and represent their relationships.* How many sheets of paper does he have on the window?" (There are 12 sheets.)

◆ Continue reading through page 16 where Jenny begins covering her window with sheets of paper. "If we look at page 17, how many sheets of paper can Jenny fit across the length of the window? The *length* is the longer side." (She can fit 6 sheets across.) Introduce the word *array* if your students are not familiar with the term, defined as a set of objects arranged in *rows* and *columns.* "How many rows of paper can she fit into her window? Rows go across the longer side of the window." (She can fit 2 rows across.) "We can make an equation to figure out the number of sheets she will use to cover the inside of the entire window by adding the number of sheets in each row. There are 2 rows and she can fit 6 sheets in each row. So the equation would be $6 + 6 = ?$ What is the answer?" (12 sheets.)

◆ "We can *reflect on the results to see if they make sense.* I will read page 17 so we can check our answer." As you read the page, verify with the students 12 sheets of paper will cover the window. "We can go back to the problem in the story now and decide whose window is bigger by comparing the total number of sheets that covered each window. It took 12 sheets of paper to cover each window. Even though the windows are a different shape, they can have the same area."

◆ Continue reading page 18 where the siblings are arguing over whose room is bigger. "I see the father on the next page and he has some newspaper in his hand. What do you think he will tell Jenny and Jeff to do with the newspaper? Think about what they did with the sheets of paper in the window." (He will tell them to cover the floor with the newspaper to see which room is bigger.)

◆ Read page 19 and show pages 20 and 21 to your students. "How can Jenny figure out the number of sheets of newspaper that will cover the whole floor when she only has newspaper taped along two of the walls? Sometimes we can *simplify a problem* to help us understand it." Show the class how you can use square inch tiles to start covering a 6 inch by 2 inch rectangular figure (see Figure 4.5).

◆ "If this is my bedroom and I start covering the floor with tiles, I can see there are 6 tiles that fit along one of my walls and 2 tiles that fit along the other wall. How can I figure out how many tiles will cover the whole floor when I only have tiles along two of the walls? I have to *identify the important quantities*, which are 6 and 2. There are 2 rows of 6 tiles. I can *represent their relationships* by making an equation, $6+6=?$ The first 6 represents the first row of tiles and the other 6 represents the second row of tiles. I know 6 plus 6 equals 12 so there will be 12 tiles that will cover the whole floor."

◆ "Now we'll go back to the book and look at Jenny's room. There are 6 sheets of newspaper that fit along one of her walls and 5 sheets that fit along the other wall. How can we figure out how many sheets will cover her whole floor? We have to *identify the important quantities*, which are 6 and 5. There are 5 rows of 6 sheets of newspaper. We can *represent their relationships* by making an equation, $6+6+6+6+6=?$ The first 6 represents the first row of newspaper, the next 6 represents the second row of newspaper, the third 6 represents the third row, the fourth 6 represents the fourth row and the fifth 6 represents the fifth row. If I go back to my

Figure 4.5

equation, **6 + 6** + 6 + 6 + 6, I already know 6 plus 6 equals 12 so there will be 12 sheets in the first 2 rows. Then I go back to the equation, 6 + 6 + **6 + 6** + 6, and see there is another 12, which is the third and fourth rows. Then there is one more 6 left in the equation, 6 + 6 + 6 + 6 + **6**, which is the fifth row. Now I have to add 12, 12 and 6."

◆ Allow time for the students to figure out the answer. They can add 12 and 12 to get 24 then count up 6 more to get 30. They can add all of the numbers by regrouping tens and ones. They can draw a picture or use manipulative tiles to add the sheets of newspaper in their visual or concrete model to get 30. If they have a calculator, explain which buttons to press so they can represent $12 + 12 + 6 =$ on their screen to get 30.

◆ "We can *reflect on the results to see if they make sense*. I will read page 20 so we can check our answer." After you read the page say, "We were right. It would take 30 sheets to cover the floor. Now I'll read another page so we can see how many sheets it will take to cover Jeff's floor."

◆ Read page 22 to your students, "How can Jeff figure out the number of sheets of newspaper that will cover the whole floor when he only has newspaper taped along two of the walls? Remember how we can *simplify a problem* to help us understand it." Show the previous 6 inch by 2 inch rectangular figure (see Figure 4.5) and prompt the students to use the same procedure to figure out how many sheets of newspaper will cover Jeff's room.

◆ "Now we'll go back to the book and look at Jeff's room. There are 6 sheets of newspaper that fit along one of his walls and 4 sheets that fit along the other wall. How can we figure out how many sheets will cover his whole floor? We have to *identify the important quantities*, which are 6 and 4. There are 4 rows of 6 sheets of newspaper. We can *represent their relationships* by making an equation, $6 + 6 + 6 + 6 = ?$ The first 6 represents the first row of newspaper, the next 6 represents the second row of newspaper, the third 6 represents the third row and the fourth 6 represent the fourth row. Before we solve this equation, can you tell if there will be more or fewer sheets that will cover Jeff's room? How *can you reflect on the results to see if they make sense?*" Allow time for your students to turn to a partner and discuss how they can answer your question by comparing the equations. (There are 5 rows of 6 sheets in Jenny's room but only 4 rows of 6 sheets in Jeff's room so it will take more sheets to cover Jenny's room.)

◆ "If I go back to my equation, **6 + 6** + 6 + 6, I already know 6 plus 6 equals 12 so there will be 12 sheets in the first 2 rows. Then I go back to the equation, 6 + 6 + **6 + 6**, and see there is another 12, which are the third and fourth rows. Now I have to add 12 + 12."

◆ Allow time for the students to figure out the answer. They can add 12 and 12 to get 24 or they may recall that 24 is the answer from this step from the previous problem. They can add the numbers by adding the tens and ones. They can draw a picture or use manipulative tiles to add the sheets of newspaper in their visual or concrete model to get 24. If they have a calculator, explain which buttons to press so they can represent 12 + 12 = on their screen to get 24.

◆ "We can *reflect on the results to see if they make sense.* I will read page 23 so we can check our answer." After you read the page say, "We were right. It would take 24 sheets to cover the floor."

◆ Read page 24 and show page 25 so your students can see that there is another area in front of the closet that is part of the area of Jeff's room. There are 2 rows of 3 sheets. Have your students represent this with a visual model by drawing the rows of sheets or with a physical model by using manipulative tiles to represent the 2 rows of 3 sheets. Then have your students write an equation to represent the 2 rows of 3 sheets.

◆ "I can see from your drawings and your tile models that you have a row of 3 and another row of 3. Use your dry erase boards to write your equation and solve it this time too." (3 + 3 = 6). "We know 24 sheets would cover the larger area of his bedroom and 6 sheets will cover the smaller area. How many sheets will cover the whole floor in his room?" Remind your students of the problem regarding the sheets of paper in Jenny's room and how the first step was adding 12 and 12 to get 24 then adding the last 6 to get 30.

◆ "We can *reflect on the results to see if they make sense.* I will read page 25 so we can check our answer." After you read the page say, "We were right. It would take 30 sheets to cover the floor. So if the area of Jeff's room is 30 and the area of Jenny's room is 30, is one of their rooms bigger?" (The area of both rooms is the same so one room is not bigger than the other.)

Students could also use inch tiles and inch graph paper to create other composite shapes. They can count all of the tiles in the composite shape or they can create an addition equation based on the number of tiles in each row of squares.

Alexander, Who Used to be Rich Last Sunday by Judith Viorst

All students can relate to the real life situation of using money. In the book *Alexander, Who Used to be Rich Last Sunday* (1978), the main character, Alexander, is complaining that his two brothers have money when he only has bus tokens. He is thinking about the previous Sunday when his grandparents visited their home and gave him and his brothers each one dollar. Throughout the story he recounts how he spent all of the money, little by little, until he had nothing left but bus tokens. (You may have to discuss what bus tokens are if students are not familiar with them.) Your students should know the coin names and values before they use this book for problem solving. They should each have a set of real coins, with at least 18 pennies, 7 nickels, 7 dimes and 3 quarters. The students will also use play dollar bills as well.

Measurement and Data 2.MD

Work with time and money.

Solve word problems involving dollar bills, quarters, dimes, nickels and pennies, using $ and ¢ symbols appropriately.

As you read the book, engage students in a discussion by asking the following questions based on the *I Can* statements:

- ◆ "When is the last time someone gave you money as a gift? Did you save it or spend it? We're going to read a book about a boy named Alexander to find out what he does with the money his grandparents give to him. You are going to use the real coins you brought to school for our money activities. We are going to keep track of the money amounts in the story to see how we *can solve problems in everyday life*."
- ◆ "We know good problem solvers *can identify important quantities and represent their relationships*. Today our quantities involve coin amounts and coin values. We will use the real coins and play dollars to keep track of the different coin amounts and values in the story."
- ◆ Read the first page then display the following word problem: Anthony has 2 dollars, 3 quarters, 1 dime, 7 nickels and 18 pennies in his room. How much money does Anthony have?
- ◆ Allow the students to use the play dollar bills and real coins to put out the number of each according to the problem. Do the same on

the document camera or Smart Board while you read the problem aloud again. "Now that each one of us has the correct amount of each dollar bill and coin, we can find out the amount of money Anthony has on the first page of the story. First we have to *identify the important quantities*. What is the value of a quarter?" (25 cents.) "The problem states Anthony has 3 quarters. How can we *represent the relationship* between the value of the quarter and the number of quarters in his room?" (We can write that 3 quarters are worth 75 cents). "What is the value of a dime?" (10 cents.) "The problem states Anthony has 1 dime. How can we *represent the relationship* between the value of the dime and the number of dimes in his room?" (We can write that 1 dime is worth 10 cents.) "What is the value of a nickel?" (5 cents.) "The problem states Anthony has 7 nickels. How can we *represent the relationship* between the value of the nickel and the number of nickels in his room?" (We can write 7 nickels are worth 35 cents.) "What is the value of a penny?" (1 cent.) "The problem states Anthony has 18 pennies. How can we *represent the relationship* between the value of the penny and the number of pennies in his room?" (We can write 18 pennies are worth 18 cents.) "Let's look at the values we wrote. How much money is there in all?" (If your students can add with regrouping, they can list all of the amounts vertically and add the ones, then the tens, to get $1.38.) "The problem also states Anthony had 2 dollars. So how much money does he have in his room?" (Anthony has $3.38.)

◆ Read the second page then display the following word problem: Nicholas has 1 dollar, 2 quarters, 5 dimes, 5 nickels and 13 pennies. How much money does he have?

◆ Allow students to use the play dollar bills and real coins to put out the number of each according to the problem. Do the same on the document camera or Smart Board while you read the problem aloud again. "Now that we have the amount of each dollar bill and coin, how can we find out the amount of money Nicholas has on the second page of the story? Let's fill out a chart to keep track of the type of bills and coins, the values of each and the total value." (See Table 4.1.)

◆ Allow students to discuss how they can add up the total value of coins in the last column of the chart and put in those values. "In which column did we *identify the important quantities* from the word problem?" (In the second column we wrote the value of the dollar and each coin. In the third column we wrote the number of each bill and coin.) "How did we *represent the relationship* between the values

Table 4.1

Type of Money	Value of Each	Total Number of Each	Total Value
Dollar	$1	1	$1
Quarter	25¢	2	50¢
Dime	10¢	5	50¢
Nickel	5¢	5	25¢
Penny	1¢	13	13¢

of the bill and the coins and the number of each bill and coin?"
(In the last column we wrote the total value of the bill and coins.)

◆ "How much money is that in all? If we look at the chart, we can
make another dollar with the 50¢ from the quarters and the 50¢
from the dimes. Then by adding the 25¢ and the 13¢, we get 38¢. So
how much money does he have in his room?" (Nicholas has $2.38.)

◆ "We *can reflect on the results to see if they make sense.* Let's look at the
visual representation in our chart to see if it makes sense that he
has $2.38. There is only 1 $1 bill, but we made another dollar with
these 2 quarters and these 5 dimes. It makes sense that we have $2.
Now we can check the nickels and pennies to see if they add up to
38¢. We can count the nickels by fives then keep counting by ones
when we get to the pennies. 5, 10, 15, 20, 25, 26, 27, 28, 29, 30 31, 32,
33, 34, 35, 36, 37, 38."

◆ "Now that we have practiced adding up coin amounts and dollar
amounts, we are going to keep track of Alexander's spending. His
grandfather gives him and his brothers each $1. Since Alexander
spends it a little at a time, we are going to first exchange the $1 bill
for the following coins—7 dimes, 4 nickels and 10 pennies. Let's
count these coins to be sure they equal $1." The students can choose
their preferred method for counting coins: counting first by tens,
then by fives then by ones; counting the like coins and recording
the amounts (70, 20 and 10), then adding those together.

◆ Once the students have verified they all have the equivalent of $1
with their coins, continue reading the book. Stop at the page where
Alexander buys gum at Pearson's Drug Store. "The last line on this
page states 'Good-bye fifteen cents.' What does that mean?"
(He spent 15¢ so it is gone.) "Use your dimes and nickels to remove
15¢ from your pile of Alexander's money. How can we use the least
amount of coins to make 15¢?" (We can use one dime and one

nickel.) "Remove one of your dimes and one of your nickels because he spent 15¢ out of his dollar so far. Put those coins on the other side of your desk. How much money does he have left right now?" Allow students to count the remaining coins and write down all answers on the board.

◆ "We *can reflect on the results to see if they make sense.* If Alexander had $1, then he spent 15¢, the amount he has left and the amount he spent should add up to $1. Let's check the answers to see which one works." (Alexander has 85¢ left.) "Tell me how you found your answer." Allow students to explain how they knew he had 85¢ left.

◆ Read the next two pages. Stop at the page where he has to pay his mom. "The last line on this page states 'Good-bye another fifteen cents.' We are going to use the same 2 coins to represent the 15¢. Remove one of your dimes and one of your nickels and put them on the other side of your desk. How much money does he have left now?" Allow students to count the remaining coins and write down all answers on the board.

◆ "We *can reflect on the results to see if they make sense.* How much did Alexander spend so far and how do you know?" (30¢ because 15¢ and 15¢ is 30¢.) "If Alexander had $1, then he spent 30¢, the amount he has left and the amount he spent should add up to $1. Let's check the answers to see which one works." (Alexander has 70¢ left.) "Tell me how you found your answer." Allow students to explain how they knew he had 70¢ left.

Continue reading the book, stopping on the pages where he spent money in order to discuss how to remove the amount using the fewest coins, counting the remaining amount of money and reflecting on the results to see if they make sense by adding up the total amount spent and their answers to see if they equal $1. As your students continue to share their strategies for counting money, emphasize the most efficient strategy and see if all of the students can count the coins that way.

What Does This Standard Mean for Grades 3–5 Problem Solvers?

Students who are successful at SMP 4 in intermediate grades can understand how word problems are connected to their everyday lives, both in school and out. They can locate the relevant quantities in word problems and are proficient in their knowledge of the situations for all four operations so they can represent those quantities. They know when to utilize the strategy of simplifying a problem in order to make better sense of the quantities and their relationships. They can look back at the original problem in light of their mathematical model to determine if the answer makes sense.

In order to fully apply SMP 4 when approaching a word problem, students in intermediate grades should be able to take ownership of their procedures by using the following *I Can* statements:

- ◆ *I can make connections between the problem and its representation in everyday life.*
- ◆ *I can identify important quantities and represent their relationships.*
- ◆ *I can use inferences and estimates to simplify the problem.*

Tiger Math by Ann Whitehead Nagda and Cindy Bickel

The book *Tiger Math* (2002) uses the true story of a Siberian tiger named T.J. born at the Denver Zoo to provide real-world examples of graphs to document T.J.'s growth, feeding habits, age and other critical data. Your students will be fascinated by the photographs of T.J. as well as learn how to read and interpret data on picture graphs, bar graphs, line graphs and circle graphs.

Measurement and Data 3.MD

Represent and interpret data.

Draw a scaled graph and a scaled bar graph to represent a data set with several categories. Solve one- and two-step "how many more" and "how many less" problems using information presented in scaled bar graphs.

As you read the book, stop at various pages in order to allow your students to engage in the following discussion, focusing on the *I Can* statements:

- ◆ On page 8, we see a graph with the title Tigers in the Wild. There are pictures of tigers in this graph so we call it a picture graph. There is a key with information in a box so we know each picture of a tiger represents 500 tigers. On the left side is a label for the information, Number of Tigers, and we can see the scale starts at 0 and goes up to 4,000 and the interval is by 500s. At the bottom of the graph there is a label for the information, Kinds of Tigers, so we know what the categories represent.
- ◆ Good problem solvers *can make connections between the problem and its representation in everyday life.* Let's think about that as we discuss the

information in these graphs. Using the graph on page 8, what can we determine about the kinds of tigers that exist in the wild if we just look at the pictures of the tigers? (There are more Bengal tigers than any other kind; there are fewer Indo-Chinese tigers and there are the fewest Sumatran tigers and Siberian tigers in the wild.)

◆ Picture graphs make it easy to summarize the relationship among the categories, which, in this case, are kinds of tigers in the wild. But with a picture graph, you are limited as to how many actual pictures you can fit into the graph. We couldn't actually fit 4,000 little tiger pictures in a graph so the author had to determine how to use one picture of a tiger to represent a large quantity. Good problem solvers *can identify important quantities and represent their relationships.* How can we determine how many more Indo-Chinese tigers there are than Siberian tigers by using the pictures and the key? (We can see there are 3 pictures of tigers in the Indo-Chinese tiger column; each picture represents 500 tigers so that is 1,500 tigers; there is 1 picture in the Siberian tiger column and that is 500 tigers; you can subtract 500 from 1,500 and the answer is 1,000.)

◆ There is often more than one way to *identify important quantities and represent their relationships.* How can we find the same answer using the scale on the left side of the graph? (We can see that the tigers in the Indo-Chinese tiger column go up to 1,500; the picture in the Siberian tiger column goes up to 500; you can subtract 500 from 1,500 and the answer is 1,000.)

◆ On page 10, the author used the exact same information from the picture graph on page 8 to create a circle graph, sometimes called a pie chart. The whole circle represents all of the tigers in the wild and each section of the circle represents the kinds of tigers. How are the two graphs different? (The circle graph does not have pictures; the circle graph uses a different color for each kind of tiger; instead of numbers in the thousands there are percentages next to each kind of tiger; this graph includes South China tigers.)

◆ One of the differences you noticed is that there is a kind of tiger on the circle graph not represented on the picture graph. Why do you think that occurred? (The section of the circle is really small; there are fewer than 500 South China tigers so they couldn't put a tiger picture on the other graph.) The authors state there are only about 40 South China tigers so they couldn't put only a little piece of a tiger picture on the picture graph. Sometimes it is better to use a circle graph because you can represent all the information about the topic.

◆ On page 14, we see another type of graph, a vertical bar graph, with the title T.J.'s Weight. Let's look at the labels, scale and other information on the graph so we *can make connections between the problem and its representation in everyday life.* How is T.J.'s weight represented and how do you know? (It is represented in pounds; there is a label on the left side of the graph that says Weight in Pounds.) How did the authors construct the scale? (They started at 0 and went up to 16; the interval is by ones.)

◆ Now let's see if we *can identify important quantities and represent their relationships.* How can we figure out how old T.J. was when he weighed 10 pounds? (We can look at the 10 on the scale on the left side and then go across to the bar that goes up to the 10 and we see that is the bar for 6 weeks.) How can we figure out how much T.J. weighed at 10 weeks? (We can look at the bars above 10 weeks and see that they go to the part of the scale which is 13 pounds.)

◆ On page 18, there is another type of graph called a line graph. This line graph displays the same information as the bar graph on page 14 but it includes information for Week 11. What do you notice about the direction of the line from Week 0 to Week 11? (The line goes up until Week 10 and then it goes down at Week 11.)

◆ It looks like T.J.'s weight went down. How *can we identify important quantities and represent their relationships* in order to figure out how much weight he lost? (We can look at the number on the scale across from the dot at Week 10, which is 13 pounds; then we can look at the number on the scale across from the dot at Week 11, which is 12 pounds; if we subtract those numbers we get 1, so he lost 1 pound.)

Continue reading the book and instead of showing your students the graph, read and display the description and data from the graph so they can construct their own graph. Remind students to include a title, to label the left side and the bottom of the graph and to determine their scale and interval based on the quantities in the data. Then show the graph on the page and have them compare their graph to the one in the book. Ask students "how many more" and "how many less" questions about their graphs. You can also challenge students to create more than one type of graph for the data, to discuss which type of graph best represents the data and to create their own "how many more" and "how many less" questions for their peers to solve.

Actual Size by Steve Jenkins

The book *Actual Size* (2004) includes illustrations on every page of various animals, or parts of animals, that are the actual size of that animal or part. It includes a description and fact about the animal and the length or weight of the animal in the appropriate unit based on size. In the back of the book, all of the animals are listed with more facts and its geographical location. Your students will be amazed at the very large, and miniature, sizes of the animals in this book.

Measurement and Data 4.MD

Represent and interpret data.

Make a line plot to display a data set of measurements in fractions of a unit. Solve problems involving addition and subtraction of fractions by using information presented in the line plots.

As you read the book, stop at various pages in order to allow your students to engage in the following discussion, focusing on the *I Can* statements:

◆ On the first page, we see the atlas moth with a 12-inch wingspan, which is shown in its actual size. Also on this page is the dwarf goby, which is the smallest fish and it is shown in its actual size of 1/3 inch in length. On the next page is the eye of the giant squid measuring 12 inches in length. That's the only part of the giant squid that can be shown in this book because its body and tentacles measure up to 59 feet long!

◆ Let's think about the length of the atlas moth and the giant squid and make a comparison about their length. We *can identify important quantities and represent their relationships* to figure out how many atlas moths would equal the same length as the giant squid. What quantities will we use in our problem? (The length of the atlas moth, which is 12 inches, and the length of the giant squid, which is 59 feet.)

◆ What do you notice about the units? (One is inches and one is feet.) We have to use the same unit in order to compare the lengths. Would it make more sense to use feet or inches in our problem and why? (It would make sense to use feet since 12 inches is the same as 1 foot.)

◆ Now how can we find our answer? Is there more than one way? (We can imagine atlas moths across the tentacles and body of the giant squid; we can draw a picture; we can divide.) Let's solve it in the way that is the most efficient. (We can divide 59 by 1 and the answer is 59 moths.)

◆ We are going to compare the lengths of a few of the animals, or their body parts, in this book. First we will represent their lengths by creating a line plot and then we'll use that information for our comparisons.

◆ I will create a chart to give you the name of the animal, description of the body part and the measurement to use in your line plot. (See Table 4.2.)

◆ In order to construct our line plot we have to determine how to create a scale and interval for the number line in our line plot. What is the smallest number represented in the data? (1/3 inch.) What is the largest number represented in the data? (12 inches.) So what should we use as the numbers on our line plot, or our scale, so we can represent all of these quantities? (We should start at 0 and go to 12.)

◆ We can see there are also some fractions so we have to be sure to represent those amounts on our line plot. What are the fractions represented in our chart? (1/3 inch and 1/2 inch.) So we should include the whole numbers from 0 to 12 and show 1/3 between the 0 and the 2 and 1/2 between the 2 and the 3. If you feel like you are running out of room or you do not want to put all of the numbers between 6 and 12, you can put a series of periods between those numbers.

◆ After you've created your number line, put an x above each measurement represented in your number line each time it appears on the chart. You should have seven x marks because there are seven animals. (See Figure 4.6.)

Table 4.2

Animal	Description	Measurement
Pygmy shrew	Body length	2 inches
Atlas moth	Wingspan	12 inches
Giant squid	Eye length	12 inches
Dwarf goby	Body length	1/3 inch
Goliath birdeater tarantula	Body length	12 inches
Great white shark	Tooth length	4 inches
Mouse lemur	Body height	2 1/2 inches

Figure 4.6

Length of animal parts

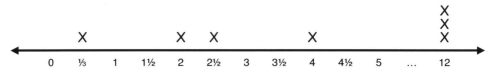

◆ Now we can use our line plot that represents the actual size of animals to *make connections between a problem and its representation in everyday life.* First we will solve some comparison problems involving the lengths of the seven animal parts represented on the line plot. In order to calculate the difference between the length of the shortest animal—dwarf goby—and the next shortest animal—pygmy shrew—we have to *identify important quantities and represent their relationships.* What are the quantities we will need in order to solve our problem? (We will use the length of the dwarf goby, which is a 1/3 inch, and the length of the pygmy shrew, which is 2 inches.)

◆ We have to decide which operation to use if we want to know the difference between their lengths. Sometimes it's difficult to determine which operation to use when one of the quantities is a fraction but we *can use inferences and estimates to solve the problem.* If the quantities were 5 and 10, what would we do to find the difference? (We would subtract 5 from 10.) So now we know we have to subtract 1/3 from 2. Is there more than one way? (We can create a common denominator; we can think about what we would add to 1/3 in order to get 2.)

◆ Now let's try another problem involving fractions. How much longer is the giant squid than the mouse lemur? Before we can solve our problem we have to *identify important quantities and represent their relationships.* What are the quantities we will need in order to solve our problem? (We will use the length of the mouse lemur, which is 2 1/2 inches, and the length of the giant squid, which is 12 inches.) This is a comparison problem. Which operation should we use for a comparison problem? (We should use subtraction.) What can we do to find the difference? Is there more than one way? (We can create a common denominator; we can think about what we would add to 2 1/2 in order to get 12.)

Your students can demonstrate and discuss how they found their answer to these problems as well as other problems involving the addition and subtraction of fractions based on information in a line plot. Your students can then go back to the book and choose other characteristics, such as the weight of animals, to create a new line plot. They can look up other facts about the animals that are not included in the book to have more practice creating, interpreting and solving problems on a line plot.

Polly's Pen Pal by Stuart J. Murphy

In the book *Polly's Pen Pal* (2005) we find the main character, Polly, sending emails to her pen pal, Ally, who lives in Canada. As they correspond, Polly notices Ally uses the metric system to measure height, weight, distance and liquid capacity. Polly's father helps her make estimates and comparisons to understand the metric measurements. At the end of the story, Polly goes on a trip to Montreal to meet Ally and is able to experience the metric system for herself.

Measurement and Data 5.MD

Convert like measurement units within a given measurement system.
Convert among different-sized standard measurement units within a given measurement system, and use these conversions in solving multi-step, real-world problems.

As you read the book, stop at various pages in order to allow your students to engage in the following activities with a meter stick and a ruler with inches and centimeters, focusing on the *I Can* statements:

◆ In our story we see Polly, who lives in New York, has a pen pal in Montreal, Canada. As they correspond and ask each other questions, Ally mentions she is 125 centimeters tall and asks Polly her height. If I asked you how tall you are, what unit would you use? (I would use feet.) If you were not exactly 4 feet tall but were halfway between 4 feet and 5 feet tall, how would you state your height? (I would say I'm 4 1/2 feet tall.) We commonly use feet as the unit to describe height and if the height is not a whole number, we use fractions to specify the exact height.

◆ As I read the story we *can make connections between the problems and their representations in everyday life.* We will convert measurement units as well as make some estimates to solve real-world problems. Let's start with Ally's height of 125 centimeters. We have a meter stick in our classroom, which is a little longer than a yardstick.

Is Ally taller or shorter than the meter stick and how do you know? (She is taller; cent means 100 so a meter is 100 centimeters.)

◆ We *can identify important quantities and represent their relationships* so let's convert Ally's height in centimeters to meters as another way to verify she is taller than the meter stick. We know there are 100 cm in 1 meter and we are converting from a smaller unit to a larger unit so we divide 125 by 100 and we will get 1.25 meters. Ally is 1 and a quarter meters tall, which is taller than 1 meter.

◆ Ally writes to Polly again and tells her she weighs 25 kilograms. Polly goes to her dad again to make some comparisons to real objects to further her understanding. We *can also use inferences and estimates to simplify a problem.* What does her father use as comparisons? (He uses a leaf for a gram and a baseball bat for a kilogram.)

◆ How many bats would we need in order to be the same weight as Ally and how do you know? (We would need 25 bats; each bat weighs 1 kilogram so if we multiply 1 by 25 we get 25 bats.) How many leaves would we need and how do you know? (We would need 25,000 leaves; kilo means 1,000 so a kilogram is 1,000 grams; we would multiply 1,000 by 25, which is 25,000.)

◆ We *can identify important quantities and represent their relationships* so we just converted Ally's weight in kilograms to grams as another way to verify the number of leaves we would need. We know there are 1,000 grams in 1 kilogram and we are converting from a larger unit to a smaller unit so we multiply 25 by 1,000 and we will get 25,000 grams.

◆ Polly and her dad are planning their trip to Montreal and want to see how far away they are from Ally. They look on a map and, using the scale, they calculate they are 450 kilometers away from Montreal. Polly asks her dad about kilometers.

◆ Let's see how we *can make connections between the problems and their representations in everyday life.* What do we have in our classroom that we can use when we think about a meter? (We can use our meter stick to think about the length of 1 meter.)

◆ Since we are familiar with a meter, let's convert 450 kilometers to meters. We *can identify important quantities and represent their relationships* so I want you to figure out which quantities to use in your calculation and whether you should multiply or divide. (We should use 450 and 1,000; we know kilo means 1,000 so there are 1,000 meters in 1 kilometer; we multiply because we want our number to be bigger than 1,000 and because we are converting from a larger unit to a smaller unit.)

◆ Yes, we should multiply 450 and 1,000 so the answer is 450,000 meters. That is a lot of meter sticks! But we can think about objects and situations as they relate to our everyday lives in order to help us make sense of a math problem.

Continue reading the story and have your students make estimates for the liters of gas her father needs to fill his tank and the distance in meters Polly has to walk to Ally's house. Then your students can discuss what they already know about measurement units to convert other distances, heights and weights to larger and smaller units, such as those of students in your class and nearby locations on a map.

Wrapping It Up

There are many ways for students to model mathematics in the word problems they encounter. They can use a physical model, visual model, number model, create a chart or table, or use written or oral procedures. By allowing students to choose how they want to represent the word problem, they can use the representation that is best for them in their development as a problem solver. They may also choose a model based on the type of word problem they encounter. When teachers use word problems with a familiar context, students can spend less time trying to understand the language in the word problem and more time making sense of the quantities and their relationships.

5

Use Mathematical Tools
SMP 5—Use Appropriate Tools Strategically

Students love to use calculators, rulers, tape measures and other mathematical tools, but do they know how and when to use them in their everyday lives? Teachers explicitly instruct their students how to use a ruler to measure an object yet students practice measuring images on a workbook page. The same is often true with calculators; teachers show their students which buttons to press to calculate an answer and then students practice pushing buttons to find answers on a workbook page. Teachers should spend additional time explicitly instructing students how to use mathematics tools when solving real-world problems.

What Does This Standard Mean for Grades K–2 Problem Solvers?

Students who are successful at SMP 5 know which types of mathematical tools are available in the classroom and how to use them. They have practiced using the tools to solve everyday problems, such as measuring space in the classroom in order to move a bookshelf or adding large numbers on a calculator to determine how many students will attend a field trip. They can determine whether mental math or a calculator would be more efficient to solve a problem.

In order to fully apply SMP 5 when approaching a word problem, students should be able to take ownership of their procedures by using the following *I Can* statements:

- ◆ *I can learn how to use different mathematical tools.*
- ◆ *I can choose the right tools to solve a problem.*

Patterns in Peru by Cindy Neuschwander

The book *Patterns in Peru* (2007) can be used to teach students how to use different math tools to solve a problem as well as describe positions of objects. In the story, twins Matt and Bibi are in Peru with their parents, who are scientists. They are in an Incan museum when they hear about the Lost City of Quwi (*coo*-ee) and set out to find it. Matt realizes he has the tunic of the secret messenger of the Lost City in his backpack, which is decorated with patterns he and Bibi use to guide their way.

Geometry K.G

Identify and describe shapes (squares, circles, triangles, rectangles, hexagons, cubes, cones, cylinders and spheres).

Describe objects in the environment using names of shapes, and describe the relative positions of these objects using terms such as *above*, *below*, *beside*, *in front of*, *behind* and *next to*.

As you read the book, engage students in discussion and activities by asking the following questions based on the *I Can* statements:

- In this story, twins Matt and Bibi are in Peru, South America with their parents, who are scientists. They hear their parents talking to Professor Herrera about a tunic, which is like a shirt, which is over 500 years old. Professor Herrera thinks the tunic belonged to a secret messenger of the Lost City.
- Matt and Bibi want to try to find the Lost City so they take off on two guanacos (gua-*na*-cos), which are like llamas. Matt gets cold and looks in his pack for something to put on. He realizes he has the ancient tunic, which their mother said has many patterns on it.
- Bibi says that the tunic could be the messenger's map. We will see how they *can learn how to use different mathematical tools.* How can they use the tunic to find the Lost City? (They can see if it shows them where to go. The patterns might be clues.)
- Now we see patterns on the tunic that have dots on lines slanting up and down. I'm going to put some circle counters on the document camera and I want you to give me directions on how to arrange the counters so they look like the circles on the tunic. I have a list of terms you can use to describe the position of objects: above, below, beside, in front of, behind, next to.

◆ Who would like to go first? (Put two circles next to each other.) Okay, I will put two of the circle counters next to each other. What else? (Put two more circles next to each other but put them below the first two circles.) That's great! I like how you used the terms *next to* and *below* to describe the position of the circles. What is the next part of the pattern? (Put two circles next to each other and put them beside the other four circles.) You're right! And you used the term *beside* so I knew where to put the two circles.

◆ Matt and Bibi find out that this pattern of four circles, then two circles shows them that first they go up the mountains on four feet, which is when they are riding their guanaco, and then they do down the mountains on their own two feet because they are very steep. So they used the pattern of two circles, four circles, two circles, four circles, to help them solve their first problem. Now Matt and Bibi are at the edge of a cliff. How can they figure out how to get over the river? (They can look at the tunic to see if there are clues to crossing the river.) Yes, Bibi finds a pattern on the tunic that shows them where to walk on the rope bridge.

◆ We *can choose the right tools to solve a problem*, just like Matt and Bibi. I have some Popsicle sticks and toothpicks to model the next part of the problem. Let's look at the lines on the tunic. I want you to give me directions on how to arrange the sticks so they look like the lines and X's on the tunic. Remember to use the list of terms to help you describe the position of objects.

◆ Who would like to go first? (Put 12 Popsicle sticks *next to* each other.) Okay, I will do that. What should I do with the toothpicks? (Take two toothpicks and cross them to make an X. Put those toothpicks *in front of* the third Popsicle stick so it looks like it is crossed out.) Let's see ... yes, there is an X on the third line on the tunic so I will put the toothpicks on every third Popsicle stick. (See Figure 5.1.)

Figure 5.1

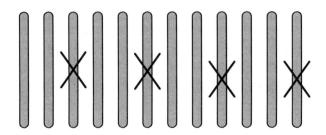

◆ We will see another example of how they *can learn how to use different mathematical tools.* How can they use the tunic to find their way across the bridge? (They knew from the pattern on the tunic that they had to step over the ropes that were crossed out.) Now that they've crossed the river, they find a large carving on a wall but they don't know how to get inside. What can they do to find how to get past the wall? (They can look at the tunic to see if there is another pattern that can be a clue to get inside the wall.)

◆ Yes, they use the tunic again as a tool by looking at the positional pattern on the tunic to figure out how they should turn the figure on the carving to get inside. I am going to show you the illustration in the book where Matt is climbing the wall to get past the carving. How can you describe Matt, the dog and the carving in the illustration using the list of terms to help you describe the position of objects? (Matt is *in front of* the carving on the wall. The dog is *below* the carving.) Why do you think Matt and Bibi are trying to open the carving? Be sure to use one of the positional terms in your answer. (They think the Lost City is *behind* the carving.)

◆ Now Matt and Bibi find themselves *in front of* another wall but this one has carvings of animals on it. How *can they choose the right tools to solve a problem?* (They can look at the tunic to see if there are clues to get inside the wall.) Yes, Bibi finds a part of the tunic that tells them how to figure out which animal carvings to touch to get *behind* the wall. But Matt reminds her that some of the tunic was torn off. What does Bibi tell him about patterns? (Patterns are predictable so if they can figure out the first part, they can figure out the rest of the pattern.)

◆ We can make a t-chart like the one Bibi makes to see how the numbers get larger for each of the nine panels on the wall. I am going to give you a highlighter and a Hundreds Chart so you *can find a pattern in the problem* and tell me which numbers to write on the t-chart. Let's look at the part of the tunic with the groups of animals. There are three foxes, six llamas and nine parrots. On your Hundreds chart, color in the 3, the 6 and the 9 while I write them in the t-chart.

◆ Now color in the 12, 15 and 18. Can you predict which numbers you will color in next? How many numbers have you colored in? (Six.) Yes, you have colored in six numbers so you have to color in three more numbers. Let's see if we can predict which numbers should be colored in next by looking at the numbers we have already colored in. What is the pattern? (We color every three numbers.) Yes, you should be coloring in every three numbers. So which numbers are next? (21, 24, 27.) Go ahead and color in the 21, 24 and 27 in the growing pattern on your Hundreds Chart while I add them to the t-chart. (See Table 5.1.) For homework you can color in the rest of the Hundreds Chart with the pattern.

Finish the story and review all of the ways Matt and Bibi used patterns from the tunic to help them solve the problems they faced as they looked for the Lost City. You can have your students practice using the positional terms with other objects, such as pattern blocks. Give students a basket of pattern blocks and a file folder to put between the partners. One student creates a pattern or design with the blocks, then uses the positional terms to describe where they placed the blocks to see if their partner can produce the same pattern or design.

How Big is a Foot? by Rolf Myller

This book *How Big is a Foot?* (1962) can be used to explore the need for standardizing the measurement of a foot. The King wants to get his wife, the Queen, a very special birthday present but she already owns everything. He decides to have a bed made for her because beds had not been invented so she did not own a bed. He speaks to his Prime Minister, who speaks to the Chief Carpenter, who speaks to the apprentice about how to make the bed. The King uses his own large feet to measure the Queen and gives the dimensions to the Prime Minister. But as the dimensions are conveyed to the apprentice, who is much smaller than the King, he uses his

Table 5.1

Panel	Number of Animals
1st	3
2nd	6
3rd	9
4th	12
5th	15
6th	18
7th	21
8th	24
9th	27

own tiny feet to measure for the bed so it turns out to be too small. Your students can practice measuring with their feet and other nonstandard units by laying them end to end.

Measurement and Data 1.MD

Measure lengths indirectly and by iterating length units.

Express the length of an object as a whole number of length units, by laying multiple copies of a shorter object (the length unit) end to end; understand that the length measurement of an object is the number of same-size length units that span it with no gaps or overlaps. *Limit to contexts where the object being measured is spanned by a whole number of length units with no gaps or overlaps.*

As you read the book, engage students in discussion and activities by asking the following questions based on the *I Can* statements:

◆ Imagine if you had to buy a birthday gift for a Queen who already owns everything. In this book, the King wants to give his wife, the Queen, a very special gift, something she does not already own. Let's find out what he decided to give to her.

◆ So we find out that the King wants to give the Queen a bed. Why is that a good birthday gift for the Queen? (Because she doesn't have a bed; beds were not invented yet; she would be the first person to ever have a bed, etc.)

◆ Since beds were not yet invented, the King had to have a bed made. Who are all of the people involved in the process of making this bed? (The King, Prime Minister, Chief Carpenter, apprentice and the Queen.) What do you notice about the size of these men in the book? (The King is big and the other men are shorter; the apprentice is small.)

◆ What question did the apprentice need answered before he could start making the bed? (He needed to know how big to make the bed.) If no one had ever made a bed, he would need to know the size of the bed. Since the bed is for the Queen, what do you think the King will have to do to answer the apprentice's question? (The King will have to find out how big to make the bed; the King will measure the Queen.)

◆ As we learn how to measure objects we *can learn how to use different mathematical tools.* If you had to measure a person, what tools could you use? (We could use a ruler, a yardstick, a tape measure, etc.) But in this story, rulers, tape measures and yardsticks have not been invented yet. So how could the King measure the Queen? (He could see how tall she is compared to him; he could see how much rope or string is the same size as the Queen; he could trace the Queen on a big piece of paper, etc.) Great ideas! We know we *can choose the right tools to solve a problem* and the King must do the same but with the tools he has in his kingdom. Let's see how he measures the Queen.

◆ Show the page where the King has the Queen lay on the ground in her pajamas and her crown, because she sometimes wore it to sleep.

◆ Turn to your partner and explain what you think the King did to measure the Queen. (He walked around her to see how tall she is and how wide she is.)

◆ What did the King use as a tool to measure the Queen? (He used his feet.) How does he know how long to make the bed? (It took six of his feet to walk on the side of the Queen.) How does he know how wide to make the bed? (It took three of his feet to walk across the bottom of the Queen.)

◆ Look at the picture of the King's feet as he walked around the Queen. He made sure there were no spaces between his feet across the bottom and again when he walked along the side of the Queen. He also didn't overlap his feet but made sure they were end to end. Let's take a few steps like the King did, walking heel to toe, heel to toe.

◆ So it seems like the problem is solved. How big should the bed be to fit the Queen? (It should be three feet wide and six feet long.) If there are no rulers, tape measures or yardsticks, how will the apprentice know what three feet wide and six feet long looks like? (He can ask the King how big to make the feet; he can use his own feet.)

◆ Show the page where the apprentice uses his own feet to measure three feet wide and six feet long.

◆ Does it look like the apprentice made sure there were no spaces between his feet when he measured? (Yes.) Does it look like he walked heel to toe so his feet were end to end? (Yes.) Do you think he will be able to make the bed the correct size so it will be big enough for the Queen? Discuss this with your partner.

◆ Good problem solvers *can use estimation to solve a problem or check the answer.* If we look at the page where the apprentice presents the bed to the King, does it look like it is the correct size? (No, it looks too small to fit the Queen.) We can see the apprentice made the bed too small for the Queen and was thrown in jail! The question here asks, "Why was the bed too small for the Queen?" You can write a few sentences or draw a picture to answer why the bed was too small for the Queen.

◆ What details did you have to think about or use to solve the problem? (We had to think about the size of the apprentice's feet and the size of the King's feet; we had to use details from the other parts of the book, like the size of the apprentice, who was much smaller than the King; we had to think about how to measure without using the same object etc.)

◆ It looks like the apprentice is trying to figure out the same problem while he sits in the jail. He realizes he needs to know the size of the King's foot so the bed can be "three King's feet wide and six King's feet long." The King is very busy so the apprentice will *have to choose the right tool to solve his problem.* What can he do to find out the size of the King's foot so he can always make things for the King? (He can trace the King's foot; he can use string or rope and cut it the same size as the King's foot, etc.)

◆ Well, they ask the sculptor to make a copy of the King's foot and give it to the apprentice and now he can use the copy as a tool to make the bed. So in this kingdom, they will always use the King's foot whenever they measure anything. That is why we call a ruler a foot, which is exactly 12 inches long. We will always know the answer to the question, "How big is a foot?"

Your students can measure the length of various objects with their feet and see if every student in the class arrives at the same answer. If you measure the same length with your feet, will the number of "feet" be more or less than their answer? Are there other body parts they can use to measure and compare, such as their pinky finger or their leg? Introduce your students to other ways people have historically used their body as a tool for measurement, such as paces for distance and hand spans for length.

Sir Cumference and the Sword in the Cone by Cindy Neuschwander

The book *Sir Cumference and the Sword in the Cone* (2003) can be used to introduce or reinforce the names and attributes of solid geometric shapes. The characters, Radius and his friend Vertex, are on a quest to find the hidden sword, which is the symbol of the next King. They must follow a puzzle and set of clues printed on parchment to locate the sword. The clues involve edges, points, faces and bases of solid figures. When using this book have the following geometric solids on display: cone, cylinder, cube, square-base pyramid, rectangular prism and triangular prism. It is suggested to have a few other sets of these solids in order to pass around to groups of students.

Geometry 2.G

Reason with shapes and their attributes.

Recognize and draw shapes having specified attributes, such as a given number of angles or a given number of equal faces.

As you read the book, engage students in discussion and activities by asking the following questions based on the *I Can* statements:

- ◆ When Radius and Vertex take the parchment to the carpenters' workshop, they find out the drawings are solid shapes flattened out which can be cut and folded to make shapes with height, width and length. Let's look at the six geometric solids I have here. Which one is the same solid shape as the first one shown on page 11? (It is a square-base pyramid.) Yes, it is the pyramid. There are several types of pyramids with different shapes at the bottom, or *base*. This one has a square as the base. There is another solid shape shown on page 11. What is its name? (It is a cube.) Yes, it is a cube, which has six *faces*. Each face is in the shape of a square. What do you think of when you hear the word *face*? (A person's face.) In geometry, a face is the flat side of a geometric solid.
- ◆ You *can learn how to use different mathematical tools* when you are solving problems if you know the names of the objects and their attributes. Radius and Vertex must keep track of the solids and their attributes as they try to solve the puzzle. They have to understand how to locate the faces, edges and points of the solids.
- ◆ I am going to have you sit in groups while I continue to read the story so I can put a set of these solids in the middle of your group. Spread them out and then each take the solid closest to you. Look for the flat parts which are the faces. How many faces are on the square-base pyramid? (5 faces.) How many faces are on the triangular prism? (5 faces.) How many faces are on the rectangular prism? (6 faces.) How many faces are on the cube? (6 faces.)
- ◆ Radius and Vertex made a chart of the solids with the faces, points and edges, as well as the calculations that match the puzzle.

(See Table 5.2.) Let's look at our solids again. How many points are on the square-base pyramid? (5 points.) How many points are on the triangular prism? (6 points.) How many points are on the rectangular prism? (8 points.) How many points are on the cube? (8 points.)

◆ They also had to figure out the number of edges for each of the solids. Let's look at our solids and feel the edges. How many edges are on the square-base pyramid? (8 edges.) How many edges are on the triangular prism? (9 edges.) How many edges are on the rectangular prism? (12 edges.) How many edges are on the cube? (12 edges.)

◆ After they create a chart to keep track of the shapes that "make two," they read on and find another clue, "The shapes that make two will pass the test; But one that does NOT must be your quest." Let's look at the chart to see which geometric solids are not listed. (The cone and the cylinder.) Yes, the cone and the cylinder are not listed so one of those solids must be their quest. Now that we know this information we *can choose the right tools to solve the problem.*

◆ Since they know they must only look at cones and cylinders in the castle, they begin searching the towers which are cylinders with cone-shaped roofs. On their search they climb up to the roof and look at the castle courtyard from above and notice circular stones in a path above the tunnel they just went through. Let's look at our solid shapes again. Which solids have a face that is the shape of a circle? (The cone and the cylinder.) They go back into the tunnel and see cone-shaped stones coming down from the ceiling.

◆ Again we have information to help us *choose the right tools to solve the problem.* Let's look at another clue Radius and Vertex use to help them decide which cone they should dig up: "Three times as tall as its base is wide; The true King's future lies inside." How should they use this information to determine which cone might contain the sword? (They could measure the width of each circle and then

Table 5.2

Geometric Solid	Faces	Points	Faces + Points	Edges	Faces + Points – Edges
Square-base pyramid	5	5	10	8	2
Triangular prism	5	6	11	9	2
Rectangular prism	6	8	14	12	2
Cube	6	8	14	12	2

if they dig it up, they would measure the height of the cone and see if the height is three times the width of the circle.)

◆ So they are going to need some tools to continue their search. What tools will they need? (They need something to measure the width and height of the cones and something to dig up the cones.) Yes, they need a measuring tape or ruler and shovels.

◆ On page 22, they realize it will take too long to dig up every cone before the sun rises. What do they know about the length of the sword? (They know the sword is about 48 inches long.) Yes, so if the sword is 48 inches, the cone should be 48 inches tall. How can they figure out how wide the cone base should be and how do you know? (They can divide 48 by 3; the puzzle said the height should be three times as tall as the base is wide.)

◆ We see Vertex has found a cone with a base 17 inches wide. What would the height of the cone have to be to fit the clue in the puzzle? (It would have to be 51 inches high because 3×17 is 51.) It looks like they found the sword in that cone, which means Vertex will be the next King!

As a follow-up activity, create word problems based on the number and shape of the faces on the solids with the answer to each problem as one of the geometric solids. As a challenge, have students create their own word problems based on the solids for homework. At the end of the book is an explanation of Euler's Law, which appeared in the story as the "two's test" and is shown in Table 5.2. Students can find other polyhedrons and apply Euler's Law.

What Does This Standard Mean for Grades 3–5 Problem Solvers?

Students who are successful at SMP 5 in the intermediate grades are familiar with various types of mathematical tools as well as with the functions of each tool. They have used tools to solve everyday problems, such as using a ruler and graph paper to determine the area of a room or using a calculator to determine how many lunches would be needed if every student ordered a hot lunch all week. They are also familiar with how tools, such as graphs, maps or concrete models, can be used both within and outside of mathematical problem solving.

In order to fully apply SMP 5 when approaching a word problem, students in intermediate grades should be able to take ownership of their procedures by using the following *I Can* statements:

◆ *I can determine when it is appropriate to use specific mathematical tools.*
◆ *I can identify relevant mathematical resources to pose or solve the problem.*
◆ *I can recognize strengths and limitations of various tools.*

Inchworm and a Half by Elinor Pinczes

In the book *Inchworm and a Half* (2001), the inchworm measures the vegetables in the garden, using her 1-inch long body as the measurement tool. One day she measures a vegetable and realizes there is more of the plant to measure, but it is less than an inch. Another worm, who is one-half inch long, explains that he is a fraction and helps her with the rest of her measurements. As the story continues there are two more worms, who represent one-third and one-fourth inch, who help the inchworm measure everything in the garden. Your students can use fractions of an inch with their ruler as a tool for problem solving.

Measurement and Data 3.MD

Represent and interpret data.
Generate measurement data by measuring lengths using rulers marked with halves and fourths of an inch. Show the data by making a line plot, where the horizontal scale is marked off in appropriate units—whole numbers, halves or quarters.

As you read the book, engage students in discussion and activities with their ruler by asking the following questions based on the *I Can* statements:

◆ We can see how much the inchworm loves to measure foods she eats, such as 2-inch peppers, 3-inch beans and 1-inch greens for salads. She makes loops to measure the objects and each one of her loops is 1 inch. But one day she realizes some of the objects she is measuring are not exactly 1 inch, or 2 inches, or 3 inches long. They are between a whole inch and the next inch. What do we call numbers that represent measurements in between the whole numbers? (They are called fractions.)
◆ We *can determine when it is appropriate to use specific mathematical tools.* The little worm who helps out the inchworm says he is a one-half inch fraction. How can he help the inchworm with her problem? (He can measure the rest of the plant if it is not another inch long; he can help her measure plants that are less than 1 inch.) So she needs an additional tool in order to measure the foods.

◆ We *can recognize strengths and limitations of various tools.* What is the strength of having a ruler with inches marked on it? (We can measure things that are exactly 1 inch, 2 inches, etc.) What is the limitation of only having inches marked? (We can't measure things less than 1 inch or if the length is between one of the inches on the ruler.)

◆ On our rulers we have inches marked on it but there are other marks between the inches. Let's look at our ruler and locate the halves. We can see the strength of having the half-inch marks on a ruler. In our story, the two worms are able to measure 7-inch celery but also 4-and-a-half-inch asparagus. The little worm makes one-half inch loops. How many loops does he have to make to be the same length as the inchworm's loop and how do you know? (He has to make two loops to be the same as the inchworm's loop; one-half inch plus one-half inch are the same as 1 inch.)

◆ What is the trouble they are having when they try to measure a carrot? (The extra length is shorter than the one-half inchworm.) So now another little worm, a one-third inchworm, helps them and they are able to measure a 9-inch cabbage but also 1-and-a-third-inch berries. The little worm makes one-third inch loops. How many loops does she have to make to be the same length as the inchworm's loop and how do you know? (She has to make three loops to be the same as the inchworm's loop; one-third inch plus one-third inch plus one-third inch are the same as 1 inch.)

◆ We can see they are now having trouble when they try to measure a tomato. So another little worm, a one-fourth inchworm, helps them. The little worm makes one-fourth inch loops. How many loops does he have to make to be the same length as the inchworm's loop and how do you know? (He has to make four loops to be the same as the inchworm's loop; one-fourth inch plus one-fourth inch plus one-fourth inch plus one-fourth inch are the same as 1 inch.)

◆ Let's look at our ruler and locate the fourths. We *can recognize strengths and limitations of various tools.* What is the strength of having a ruler with inches, halves and fourths marked on it? (We can measure objects that are a whole inch but also objects that are halfway or one-fourth of the way between the inches.)

Your students can practice measuring objects in their desk, in the room, around the school or at home to the nearest inch, half inch and quarter inch. Then they could list the objects in a chart and make a line plot with their data.

Sir Cumference and the Great Knight of Angleland by Cindy Neuschwander

In this book, *Sir Cumference and the Great Knight of Angleland* (2001), we witness another adventure from Radius, the son of Sir Cumference and Lady Di of Ameter. Radius and his horse set off to find King Lell, who has disappeared. His parents give him a medallion, which is actually a protractor, as a good luck charm. Radius arrives at a giant castle and must decipher a riddle in order to get to the King. He finds his medallion useful as he measures right, straight, obtuse and acute angles in order to choose the correct path. Your students will learn about these types of angles and their degree measures as well as how to use a protractor as a mathematical tool.

Measurement and Data 4.MD

Geometric measurement: understand concepts of angle and measure angles.

Recognize angles as geometric shapes that are formed whenever two rays share a common endpoint, and understand concepts of angle measurement.

As you read the book, engage students in discussion and activities by asking the following questions and activities using a protractor, based on the *I Can* statements:

◆ On pages 4 and 5, we see Radius and his horse practicing in the riding ring following orders from his instructor, Sir D'Grees. Radius is told to ride a "knightly right angle" in the ring by having his horse trot in a straight line from the edge of the ring to the center and then turn in an exact right angle which forms a corner. Then Sir D'Grees tells Radius to "double the right angle to make a straight angle."

◆ In order to become a knight, Radius must go on a quest to save someone in need of help. The family learns that King Lell is missing and send Radius in search of him. On page 8, we see Sir Cumference has given his son a medallion to take on the quest, not knowing what the numbers around the edge of the medallion signify. Does the medallion look like a mathematical tool you have seen? (It looks like a protractor.)

◆ Good problem solvers *can determine when it is appropriate to use specific mathematical tools.* Let's look at our mathematical tool, a protractor, and practice using it throughout the story. When Radius was in the riding ring, which is the shape of a circle, he rode to the center of the circle then turned 90 degrees and headed in another direction to leave the circle. (See Figure 5.2.) The center of the circle becomes the vertex and the two lines, riding in to the center then back out again, are called rays. When two rays have a common endpoint they form an angle.

◆ A protractor is a tool we can use to measure an existing angle or to create an angle of a specified degree measure. Sir D'Grees told Radius to make a right angle, which is 90 degrees. Then he told Radius to double the right angle. What would the new degree measure be if we double the right angle? (It would be 180 degrees.) Yes, a 180-degree angle is called a straight angle because it is a straight line. When Radius rode his horse to create a straight angle, all he had to do was ride from one end of the ring to the other through the center, which created a diameter of the circle. (See Figure 5.3.)

Figure 5.2

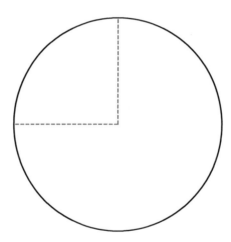

Figure 5.3

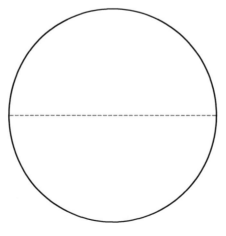

◆ On page 10, Radius reaches a village with cottages that have pointed rooftops in steep angles. He makes a comment that the village is cute. Then he rides through the Mountains of Obtuse to get to a castle where he believes King Lell is hidden. When he arrives at the castle he finds a parchment hanging on the door with a warning. On his journey there are some words we will recall later: cute, obtuse, knightly right, big, straight, slight.

◆ We *can identify relevant mathematical resources to pose or solve a problem.* This parchment and its words serve as a resource, though Radius does not fully understand the significance of the words when he first reads through the warning. On page 15, we see the stone floor of the room creates a circular pattern that reminds him of his riding ring. He uses his medallion to interpret the part of the warning that states, "Make a Knightly right," as a 90-degree angle so he knows which direction to travel to continue his search.

◆ We can also *recognize strengths and limitations of various tools.* Radius is certainly realizing the strengths of his medallion and the purpose of the numbers on the edges, which was unknown to his parents. On our protractor we see two sets of ruler markings starting at 0 on each end and ending at 180, with 90 at the top of the arc. Just like Radius, we have to learn which set of markings to read so we can be accurate with our angle measures.

◆ On page 16, Radius reads the next part of the warning, "Finding next Big, Straight, and Slight." He thinks back to the Mountains of Obtuse and the shape of the mountains with the peak at the top and then spreading out in opposite directions like big angles. He approaches a section of the castle with several hallways. Each hallway has a circle carved in front of it with a design resembling the shape of his medallion and an arrow along the upper edge of the circle. As he holds the medallion over each circle he notices the arrow lines up with numbers on the outer edge of his medallion. He figures out one of the circles has a 55-degree angle and others have angles less than 90 degrees, except for one measuring 120 degrees.

◆ Let's look at our protractor closely. How did Radius decide which set of numbers to use to measure each angle? Think about the Knightly right angle! (He used the smaller numbers if the angle was less than 90 degrees and the larger numbers if the angle was more than 90 degrees.) That's right, we have to always think about the right angle then decide if we are measuring, or creating, an angle that is 90 degrees, greater or less than 90 degrees. Later we'll find out the names of angles which are greater and less than 90 degrees!

◆ Radius enters the hallway with the circle measuring 120 degrees because it is bigger, or greater, than 90 degrees. He is now under the castle and encounters two bridges, each crossing a fire pit at different angles. What was the next clue in the warning and how do you think it would affect his decision? (The next clue is Straight; he should look for a straight angle, which is 180 degrees.)

◆ It worked! He chooses the bridge that is straight and gets safely past the fire pit. On page 21, we see Radius at the end of a tunnel with seven other tunnels from which to choose. The floor is again in the shape of a circle and the image on the floor in the center resembles his medallion. What was the next clue in the warning and how do you think it would affect his decision? (The next clue is Slight; he should look for a slight angle, which is less than 90 degrees.)

◆ Yes, he thinks about the slight angles in the rooftops of the cute little village he passed through at the beginning of his quest. He uses his medallion on the floor again and finds the smallest angle is 40 degrees and enters that tunnel.

◆ At the end of the tunnel he arrives at four corridors. What was the next clue in the warning and how do you think it would affect his decision? (The next clue is Right to reach the King; he should look for a right angle, which is 90 degrees.)

◆ Yes, but if we look at pages 22 and 23 we see four circles with angles marked and they all appear to be right angles. Remember, we *can recognize strengths and limitations of various tools*. Radius uses his medallion, which has the strength to show him the exact angle measure, but he measures them so quickly they all seem to be 90 degrees. When he finally slows down and measures once more he realizes one is 93, one is 85, one is 89 and only one is 90 degrees.

◆ We see he chose the correct corridor and found King Lell behind the door. Once everyone finds out Radius saved the King, they all visit the castle and King Lell proclaims Radius a knight. Everyone wants to know how Radius found his way through the maze of the castle.

◆ He tells them the purpose of the numbers around the edge of the medallion and how they helped him measure angles. He called them *degrees* in honor of his teacher, Sir D'Grees. He also refers to angles less than 90 degrees and calls them *acute* angles based on the narrow rooftop angles he saw in the cute little village. The angles that are bigger than 90 degrees he calls *obtuse* based on the big Mountains of Obtuse.

Your students can practice measuring angles in the book with the medallion attached to the back of the book as they enjoy re-reading the story on their own or with a peer. They can also create a new adventure for Knight Radius in which he faces decisions based on the type of angles and precise angle measures in order to fulfill his quest.

Perimeter, Area and Volume by David A. Adler

In the book *Perimeter, Area and Volume* (2012), monsters are the characters who provide examples and definitions of two- and three-dimensional objects and the terms length, width and height. Then these terms are used to calculate perimeter and volume as well as to introduce cubic units. Your students can make connections between the examples from the book and real-world objects and experiences involving tools for measurement.

Measurement and Data 5.MD

Geometric measurement: understand concepts of volume and relate volume to multiplication and to addition.
Recognize volume as an attribute of solid figures and understand concepts of volume measurement. A cube with side length 1 unit, called a "unit cube," is said to have "one cubic unit" of volume, and can be used to measure volume.

As you read the book, engage students in discussion and activities by asking the following questions based on the *I Can* statements:

- ◆ Has anyone been to a movie shown in 3D? How was it different than a regular movie? (It looks like things are coming out at you; it seems more real because it seems like you could touch the things in the movie.) Yes, the objects on the screen not only have height and width, they have depth. Those are called dimensions. A movie in 2D means it only has two dimensions—height and width. A movie in 3D has three dimensions—height, width and depth.
- ◆ In our book, we can see the monsters in the story being described with words such as tall, short, wide, narrow, fat and thin. We *can determine when it is appropriate to use specific mathematical tools.* If we want to know exactly how big something is we have to know more specific information, such as the dimensions of an object, and we can use various tools to measure them.
- ◆ For example, if we look at our book at the pictures of two fenced-in yards, we can estimate which one looks bigger. But if we want to

know for sure we have to measure the dimensions of length and width and then calculate the perimeter. Let's quickly review perimeter. What is the perimeter of Yard 1 if two sides are 80 feet, one side is 60 feet and the other side is 66 feet, and how do you know? (The perimeter is 286 feet; you add all of the sides; you double the side that is 80 feet, then add the other sides.)

◆ What is the perimeter of Yard 2 if two sides are 90 feet, one side is 40 feet and the other side is 42 feet, and how do you know? (The perimeter is 262 feet; you add all of the sides; you double the side that is 90 feet, then add the other sides.) So we know for sure that Yard 1 is bigger.

◆ On the next page, we see a circle with a radius of 25 inches. We will review circumference but first we can think about how we *can recognize strengths and limitations of various tools.* What are some choices of tools we could use to help us with our calculation? (We can use a calculator, a computer or paper and pencil.) What are the strengths and limitations of each of these tools in regard to this problem? (The strength of the calculator and computer would be they can multiply $25 \times 2 \times$ pi more quickly than we can do it on paper and they can calculate pi to more places or we can use the pi key; the limitations of the paper and pencil are we have to do the multiplication by hand and it takes longer and we have to use 3.14 and not a pi key or not write pi to as many places.)

◆ What is the circumference of this circle and how do you know? (The circumference is 50Π or about 157 inches; we use the formula for circumference $2\Pi r$.) So we used our calculator, which was an appropriate tool.

◆ Let's review one more formula for two-dimensional objects. The movie screen in the book is 12 feet wide and 16 feet long. What is the area and how do you know? (The area is 192 square feet; the formula is length times width.) What tool did you use to find out the answer? (I used paper and pencil; a calculator.)

◆ What if we didn't know the formula for circumference of a circle or area of a rectangle or didn't know how to calculate the perimeter of a polygon? (We could look it up in our math book or do a search on the computer; we could ask someone.) So if you don't know the answer you *can identify relevant mathematical resources to pose or solve a problem.*

◆ Now we'll move on to three-dimensional objects. If we look at our book at the pictures of the Jumbo and Large popcorn boxes, we can make an inference that the Jumbo box is bigger and holds more popcorn—that's why they call it Jumbo! But if we want to know how much bigger the Jumbo box is than the Large box, or how much more popcorn it can hold, we have to calculate the *volume*.

◆ We *can determine when it is appropriate to use specific mathematical tools.* If we want to calculate the volume we have to know the dimensions of each popcorn box. What tools would we use to measure the three dimensions? (We could use a ruler or tape measure.) We can see the monster in the illustration using a ruler to measure the height, width and depth of each box. The formula for volume is length × width × height with all three measurements in the same unit of measure.

◆ We will use the formula to find the volume of each box but first we can think about how we *can recognize strengths and limitations of various tools.* What are some choices of tools we could use to help us with our calculation? (We can use a calculator, a computer or paper and pencil.) What are the strengths and limitations of each of these tools in regard to this problem? (The strength of the calculator and computer would be they can multiply the numbers quickly; the limitations of the paper and pencil are that we have to do the multiplication by hand and it takes longer.)

◆ Let's have half of the class use their calculator and the other half use paper and pencil and see which seems to be the better tool for this problem:

How much larger is the Jumbo box than the Large box?
Jumbo box: 8 feet high, 6 feet wide, 4 feet deep
Large box: 6 feet high, 4 feet wide, 3 feet deep

◆ Although most of you predicted the calculator would be the most efficient tool, it seems many of you did the problem on paper and pencil with minimal calculations. Why was this problem better suited to the tools of paper and pencil? (The numbers were small so we could do some or all of the multiplication in our head and write the answer down; it was faster than typing in all of the numbers and symbols in a calculator.)

◆ Now we see one of the monsters holding up two cubes, one in each hand. What do you notice about the dimensions of each cube? (The big cube is 1 inch for all dimensions; the small cube is 1 centimeter for all dimensions.)

◆   These are called *unit cubes* and are used to measure volume. The big cube is a cubic inch and the smaller cube is a cubic centimeter. There are also cubic feet and cubic meters to measure volume of larger objects.

After you finish the story, set up the following word problem for students to solve using a choice of math tools, such as a calculator, unit cubes (cubic inches), ruler, paper and pencil:

How many cubic inches would fill the Jumbo popcorn box? How many cubic inches would fill the Large popcorn box? Which tools would you use and why would you use those tools? Explain how you used the tools to find your answers.

Encourage your students to bring in boxes from home and find the volume of each container using a unit cube and by measuring the length, width and depth with a ruler. They can predict which box will have the largest volume before they present their findings.

Wrapping It Up

Students should be taught to use a variety of mathematical tools to help them solve word problems when appropriate. But first, they should become familiar with the types of tools available, which type of measurement is best suited by the tool, how to use the tool, which tool is most efficient and which units are recorded in their answer. Allow students to practice using more than one tool to solve the same problem and justify which tool would be the most efficient based on their experiences.

6

Attend to Precision

SMP 6—Attend to Precision

Most teachers have experienced the frustration of asking students how they arrived at their answer only to hear, "I just knew it" or "that was the number in my calculator." Teachers may be further frustrated to realize their students do not know *how* they solved a word problem even if their answer is correct. More often than not, students simply cannot verbalize their solution process. In order to provide an explanation, a student must possess strong content knowledge, an understanding of the goal of the word problem and why the answer makes sense.

What Does This Standard Mean for Grades K–2 Problem Solvers?

Students who are successful at SMP 6 in the primary grades know how to interpret the symbols and the mathematical language in a word problem. They know the definition of mathematical terms used in their grade level as well as how to use the terms to explain their solution process. They are becoming familiar with some problem solving strategies which they can apply to the word problem after much modeling and thinking aloud demonstrated by the teacher.

In order to fully apply SMP 6 when approaching a word problem, students should be able to take ownership of their procedures by using the following *I Can* statements:

- ◆ *I can define the meaning of mathematical symbols.*
- ◆ *I can correctly label my diagram, drawings, graphs and units in the answer.*
- ◆ *I can explain how I solved a problem using mathematical terms.*

If You Were a Triangle by Marcie Aboff

Students can learn the definition of a triangle and related mathematical terms in the book *If You Were a Triangle* (2010). The characters in this book use triangles which appear in the environment, such as a Yield sign, a triangle instrument and slices of watermelon. On the left side of each two-page spread, the sentence begins with "If you were a triangle...." The terms used in the book are mathematically correct but are written in a way a child could understand, with pictures and examples used for support. You should have several examples of the following triangles (in various sizes) to display as well as smaller versions for each student or pair of students to share: isosceles triangle, right triangle, equilateral triangle, scalene triangle.

Geometry K.G

Analyze, compare, create and compose shapes.

Analyze and compare two- and three-dimensional shapes, in different sizes and orientations, using informal language to describe their similarities, differences, parts and other attributes. Compose simple shapes to form larger shapes.

As you read the book, engage students in discussion and activities by asking the following questions based on the *I Can* statements:

◆ In the same way we learn what good readers do, we can learn what good math problem solvers do. One thing good math problem solvers do is *use mathematical symbols and terms correctly.* In this book, we will learn about *triangles* and there will be many terms we will put up on our Math Word Wall. It says on page 3 that a triangle is "a flat, closed figure with three straight sides." Which figures on pages 4 and 5 are triangles? (The Yield sign; the cats' ears; shape on the front of the drum; the guitar; the triangle instrument; the slice of watermelon.) Yes, these are all figures that are *flat* and they have three sides. We can also call flat figures *two-dimensional* figures.

◆ On pages 6 and 7, we can see another important part of a triangle—it has *straight sides.* On page 8, we have a new word that we will use when we talk about some shapes. The word is *polygon.* The definition of a polygon is that it is a flat, closed figure with three or more straight sides. Triangles always have three sides so they are polygons. Shapes such as squares and rectangles that have four sides are polygons too. But polygons have straight sides so a circle is not a polygon. A *closed* figure means there can't be any openings. The sides all have to be touching another side.

◆ On page 10, we can see another important thing to remember about triangles, which is that they have three corners, which are called *angles*. If we look at the picture, we see children riding their bike on a bike trail with three sides. The corners where they have to turn are the angles. It is where two of the sides of a polygon are touching.

◆ Let's read about a right triangle, which is a type of triangle. I want you to use the set of triangles I gave you for this next set of problems. I'll read the description and I want you to hold up a triangle that matches the description. Here is the first one: two sides are *perpendicular*, which means one side is lying flat and one side would be standing straight up from it, like when the clock hands show 9:00. (They should hold up a *right triangle*.) Yes, it is a right triangle because the type of angle that is perpendicular is called a *right angle*.

◆ Here is another problem we're going to do with your set of triangles. Hold up a triangle that matches this description from the book, "Two of your sides would be the same length. Your third side would be a different length." (They should hold up an *isosceles triangle*.) Yes, it is called an isosceles triangle and it only has two sides that are the same length.

◆ Here is another problem we're going to do with your set of triangles. Hold up a triangle that matches this description from the book, "Your three sides would be the same length." (They should hold up an *equilateral triangle*.) Yes, it is called an equilateral triangle and all three sides are the same length.

◆ We can also see on pages 20 and 21 that we can put triangles together in certain ways to form other shapes. If we put two right triangles that are the same size together with their longest sides touching, we can make a rectangle. If we put two isosceles right triangles that are the same size together with their longest sides touching, we can make a square. If we put two equilateral triangles that are the same size together with two sides touching, we can make a rhombus.

After you have finished reading the book, your students can review their various triangles and glue them on a poster. This can be an opportunity for students to learn how they *can correctly label a diagram, drawings, graphs and units in the* answer by labeling each triangle with the correct name, as well as labeling the sides and angles. They can refer to the Word Wall or sit with a partner using the book. You could also provide graph paper and a straight edge for students to draw each type of triangle, labeling the sides and angles, or cardboard examples for students to trace. Create a center or homework activity with several of each type of triangle (using various colors and sizes) for students to sort into categories. Include other shapes such as circles and ovals, as well as other polygons so they can further analyze and compare similarities and differences among the figures.

Lemonade for Sale by Stuart J. Murphy

The book *Lemonade for Sale* (1998) can be used to introduce or reinforce the concept of representing and interpreting data as well as labeling a graph. The children in the story want to have a lemonade stand to earn money to repair their clubhouse. They create a bar graph to keep track of the number of cups of lemonade they sell each day for a week. As the week progresses, they fill in the bar graph and compare each day to the previous day, providing opportunities for problem solving involving the information presented in the bar graph. You can create the bar graph in the story with your students by using a large piece of graph paper or recreating it to display on a document camera.

Measurement and Data 1.MD

Represent and interpret data.

Organize, represent and interpret data with up to three categories; ask and answer questions about the total number of data points, how many in each category, and how many more or less are in one category than in another.

As you read the book, engage students in discussion and activities by asking the following questions based on the *I Can* statements:

◆ We can see the children in the story want to sell lemonade in order to earn money to fix up their clubhouse. Danny suggests keeping track of their sales. What is Sheri's idea for keeping track of the sales? (She will make a bar graph and show the number of cups they sell each day.) Let's look at the bar graph on page 7. What information is on the graph? (There are numbers on the left side and days of the week on the bottom.)

◆ We have to be sure we *can use mathematical symbols and terms correctly* when we solve word problems so we are going to look closer at the numbers and words on Sheri's graph. Which numbers are on the graph and where did she write the numbers on the graph? (The numbers are from 0 to 90 but they are only the tens. They are next to the lines that go across the graph.) The numbers on the left side of the graph are the *scale*. You can see there isn't room to fit all of the numbers from 0 to 90 so she only put the tens on the scale. She put them next to each *horizontal line* starting at the bottom, which would be 0.

◆ Now let's look at the words on the bottom of the scale. What words did she write and where did she write the words on the graph? (She wrote Mon., Tues., Wed., Thurs. and Fri. for the days of the week. She wrote them under every other column of squares on the graph.) She left space between the words so it will be easier to see the different columns once they are colored in.

◆ What do the numbers represent? (They represent the number of cups sold each day.) What will she color in if they sell 10 cups on Monday? (She will color in one square above the Mon.) What if they don't sell exactly 10? What if they sell 8? (She will color in the square but not all the way to the 10.)

◆ I am going to make the same graph as the one Sheri made so we can use it to solve problems about the cups of lemonade sold. I will put the numbers 0 through 90 on the left side of the graph, making sure I am writing only the tens on the scale and that each number is next to a horizontal line. Next I will write the abbreviations for the days of the week under the squares, leaving one column between each day.

◆ There is one more thing we have to do before we can continue reading the story. We also have to be sure we *can correctly label diagrams, drawings and units in the answer*. Sheri did not label the graph in the book but I am going to label ours. If someone looked at our graph and did not know we were reading *Lemonade for Sale*, they would not know what the numbers and days of the week represent. What can we write on the left side, next to the numbers, that can let people know what these numbers represent? (They are the number of cups of lemonade they sold.) I will write Number of Cups.

◆ Now we have to label the bottom of the graph too, where the days of the week are written. What do these days represent? (They are the days the kids are selling their lemonade.) I will write Days

Selling Lemonade. The last thing we have to add is a title. We have to let people know this graph is being used to keep track of the number of cups of lemonade sold. How about, Cups of Lemonade Sold. The title of a graph should be short and is usually not a whole sentence so we don't put a period at the end but we do capitalize the important words.

◆ Now we can start filling in the graph as we read the story and then solve some problems with the graph. On the first day of their lemonade stand Sheri said they sold 30 cups. What should we color in to show they sold 30 cups of lemonade? Be sure you can explain why. (Color in 3 boxes above the Mon. because each box is 10. Three boxes will show 30 cups sold.)

◆ On the second day of their lemonade stand Sheri said they sold 40 cups. What should we color in to show they sold 40 cups of lemonade? Be sure you can explain why. (Color in 4 boxes above the Mon. because each box is 10. Four boxes will show 40 cups sold.)

◆ Now we can use the graph to solve some problems. I'm going to show you a word problem and I want you to use our graph to solve it. Be sure to label your units and be sure you *can explain how you solved the problem using the mathematical terms.*

> The kids sold 30 cups of lemonade on Monday. They sold 40 cups of lemonade on Tuesday. Did they sell more lemonade on Monday or Tuesday? How many more?

◆ I want you to write down your answer to the first question on a dry erase board and hold it up. I see that most of you wrote Tuesday. First, we have to know how to read the bar graph. How do you know how many cups were sold on Monday and Tuesday? Explain your answer using terms such as scale, horizontal line and bar graph. (I know they sold 30 cups on Monday because there are 3 boxes colored in above the Mon. and the colored boxes go up to the horizontal line next to the number 30 on the scale. I know they sold 40 cups on Tuesday because Sheri is drawing a line from the 40 on the scale across the horizontal line to the top of the colored boxes above Tues.) Why do you think they sold more cups on Tuesday? (40 is a bigger number than 30 so they sold more on Tuesday.) Yes, they sold more on Tuesday. The second question is, how many more? (They sold 10 more cups.) There is more than one way to figure out this answer. What are some ways you know the answer is 10? (You could count up from 30 to 40; you could see that there is one more box colored in on Tuesday and each box is for 10 cups; you could make a subtraction problem which is 40−30.)

◆ I heard someone say that it could be a subtraction problem, which would be written 40−30=? This is called a comparison situation, where the difference is unknown. (See Table A1.1 in Appendix.) We could also solve it with an addition situation where the addend is unknown, which would be written 30+?=40. If you said you could solve the problem by counting up from 30 to 40, then you were using the addition situation with the addend unknown.

◆ I am going to give you another word problem that you can solve by looking at the graph on page 14 again. Show me your answer and write an addition or subtraction situation that would match the problem:

> The kids sold 30 cups of lemonade on Monday and 40 cups of lemonade on Tuesday. How many cups of lemonade did they sell both days?

◆ Show me on your dry erase board how you solved this problem. (30+40=70.) How did you know that you would add the numbers? (We had the first amount of lemonade, which was 30, and another amount, which was 40, and then we had to put the two amounts together.) Yes, this is an example of a *put together* addition situation with the total unknown.

Continue reading the story, making up word problems to go with each page that your students can solve as a class, with a partner or individually based on their skill level and familiarity with the addition and subtraction situations. They can then create their own bar graph with similar information provided (number of candy bars sold in a week), using a checklist of the components needed to correctly label their graph:

◆ Write the numbers on the left side of the graph for your scale.
◆ Label the vertical axis of the graph.
◆ Write words below each bar at the bottom of the graph.
◆ Label the horizontal axis of the graph.
◆ Write a short title for the graph at the top.

Measuring Penny by Loreen Leedy

In the book *Measuring Penny* (1997), Lisa's teacher assigns a homework project in which the students have to choose an object using standard and nonstandard units of measure. Lisa chooses to measure her dog, Penny, and other dogs in her neighborhood. She creates drawings and diagrams to keep track of her results, using various mathematical symbols and terms while she labels her units. You and your students can use examples in the book to create word problems relating addition and subtraction to length.

Measurement and Data 2.MD

Relate addition and subtraction to length.
Use addition and subtraction within 100 to solve word problems involving lengths that are given in the same units, e.g., by using drawings (such as drawings of rulers) and equations with a symbol for the unknown number to represent the problem.

As you read the book, engage students in discussion and activities by asking the following questions based on the *I Can* statements:

◆ Now that you are developing as problem solvers, you should be able to *correctly label diagrams, drawings, graphs and units in your answers.* In this book, Lisa is given a homework assignment which requires her to measure an object of her choice in as many ways as she can, record her results by including the number and unit, make at least one comparison and use both standard and nonstandard units. Her teacher, Mr. Jayson, provides examples of units they can use but he tells them to be creative. Lisa will have to *correctly label diagrams, drawings, and units in her answers.*

◆ When Lisa gets home, she gets the idea to measure her dog, Penny. She is already thinking about comparing Penny to at least three other dogs in her neighborhood and draws a picture of Penny and the other dogs, labeling each type of dog. Let's look at page 7 where she is measuring the length of each dog's nose with a ruler. I'm going to give you a word problem to solve using this picture:

> Lisa is measuring the length of the nose of some of the dogs using her ruler. The sheepdog's nose is 4 inches, Penny's nose is 1 inch and the pug's nose is half an inch. How many inches longer is the sheepdog's nose than Penny's nose?

◆ In order to solve this, let's think about whether we would add or subtract in order to find the difference in length of the noses. I am going to list the common addition and subtraction situations and I want you to think about which one fits this word problem. The situations are *add to, take from, put together, take apart* and *compare.* Which one fits this word problem? (We are going to *compare* the lengths of their noses.) In this type of comparison, we subtract the smaller length from the larger length. I want you to use your dry erase boards to write an equation with a symbol for the unknown number. You can use *l* for length.

◆ Let me see your equations. The correct equation should be $4 - 1 = l$. I know you can subtract these numbers with mental math so go ahead and erase the *l* and put in the length. (3) Did you label your unit? (3 inches.)

◆ Are dog biscuits a standard or nonstandard unit and how do you know? (They are a nonstandard unit because all dog biscuits aren't the same size; you wouldn't always know how long something is in dog biscuits.) Even though the biscuits are nonstandard units, we can still use addition and subtraction to compare length. Now I'm going to give you another word problem for the picture on page 8:

> Lisa is measuring the length of the dogs' tails using dog biscuits. The greyhound's tail is 10 dog biscuits long. The terrier's tail is 4 dog biscuits long. How many more dog biscuits would Lisa need to add to the 4 biscuits on the terrier's tail so that it would equal the 10 biscuits on the greyhound's tail?

◆ In order to solve this, let's look at the list of the common addition and subtraction situations so we can think about which one fits this word problem. The situations are *add to, take from, put together, take apart* and *compare.* Which one fits this word problem? (We are going to *compare* the number of dog biscuits that are on the dogs' tails.) This is also a comparison situation but it is different than the example with the length of the dogs' noses. They are both using length and they are both comparing the length, but how is the goal of this problem different than the other one? (This problem wants us to figure out how many biscuits would equal 10 and 10 is the big number. In the other problem we put the big number at the beginning of the equation and we subtracted 1.)

◆ I want you to use your dry erase boards to write an equation with a symbol for the unknown number. You can use *b* for biscuits.

Remember what we said about how the equation should equal 10 in this problem.

◆ Let me see your equations. The correct equation should be $4 + b = 10$. I know you can figure this out with mental math so go ahead and erase b and put in the number. (6) Did you label your unit? (6 dog biscuits.)

◆ For the first problem we wrote $4 - 1 = l$ and that is called a difference unknown type of situation because we didn't know the difference between the two amounts. In the second problem we wrote $4 + b = 10$ and that is called an addend unknown because we didn't know the other number to add that would equal 10.

◆ Let's look at another way Lisa measured Penny on pages 25 and 26. On these pages, Lisa is showing how she can measure the temperature in degrees Fahrenheit that Penny prefers when she goes on her walk. Lisa drew pictures of Penny and labeled the temperature in degrees at the bottom of each picture. I am going to give you a word problem about temperature:

> Penny likes to go for a long walk when it is 60 degrees outside. When it is 80 degrees outside, Penny thinks it is too hot for a long walk. How many degrees hotter is 80 degrees than 60 degrees?

◆ In order to solve this, let's look at the list of the common addition and subtraction situations again so we can think about which one fits this word problem. The situations are *add to, take from, put together, take apart* and *compare*. Which one fits this word problem? (We are going to *compare* 80 degrees with 60 degrees.) This is another comparison situation. Is it like the problem with the length of the dogs' noses where we wrote $4 - 1 = l$ or is it like the problem with the length of the dogs' tails where we wrote $4 + b = 10$? Is there a way we can solve this by addition or by subtraction, depending upon how we write the equation? I want you to turn to your partner and talk about how you could create an equation using the addend unknown and an equation using the difference unknown. Remember, we *can explain how we solved a problem using mathematical terms*.

◆ I want you to use your dry erase boards to write the equations you and your partner discussed, using a symbol for the unknown number. You can use t for temperature. It is okay if you and your friend only created one equation so far.

◆ The equation for the difference unknown is the most common equation to use for this type of problem. The equation should be $80-60=t$. If you wrote an equation using the addend unknown, it would be $60+t=80$. In fact, when you are doing your subtraction for $80-60$, you might be thinking about what the missing addend would be. For example, I know it is efficient to think about addition problems when trying to solve subtraction problems. For example, I know $10-5=5$ because I know $5+5=10$.

◆ It doesn't matter which equation you and your friend are using to solve this word problem because the answer would be the same. What is $80-60$? (20) What number would I add to 60 so that it equals 80? (20)

You and your students can use many of the other pages to create word problems that can be solved by drawing a picture or diagram or writing an equation. Students should be able to see multiple ways of solving the same problem from the story and practice using different units in the story, such as centimeters, feet and cups.

What Does This Standard Mean for Grades 3–5 Problem Solvers?

Students who are successful at SMP 6 in the intermediate grades can define and use mathematical terms appropriately at their grade level and consistently use these terms to explain their solution process. They are familiar with many problem solving strategies and many types of word problems, including multi-step problems. They know when to apply a certain strategy to the word problem and will correctly label drawings, diagrams and units in their solution.

In order to fully apply SMP 6 when approaching a word problem, students in intermediate grades should be able to take ownership of their procedures by using the following *I Can* statements:

◆ *I can apply definitions and state the meaning of symbols when I communicate my answer.*
◆ *I can correctly apply labels and specify units in my answer.*
◆ *I can use precise language to explain how I solved the problem.*

Chimp Math by Ann Whitehead Nagda and Cindy Bickel

In the book *Chimp Math: Learning about Time from a Baby Chimpanzee* (2002), the reader observes the daily routine of Jiggs, the baby chimpanzee raised in a Kansas zoo and then in Denver zoo. The workers at the zoo had to carefully monitor the chimpanzee's feeding schedule, weight, physical activity and progress with skills such as learning sign language. Your students will see Jiggs' life measured using timelines, charts, clocks and calendars.

Measurement and Data 3.MD

Solve problems involving measurement and estimation of intervals of time.

Tell and write time to the nearest minute and measure time intervals in minutes. Solve word problems involving addition and subtraction of time intervals in minutes.

As you read the book, engage students in discussion and activities by asking the following questions based on the *I Can* statements:

- ◆ We're going to be reading a book about a baby chimpanzee who is raised in the Denver zoo because his mother did not take care of him after he was born. There are going to be many concepts of time in this book, such as the timeline shown on page 8. We can see some years on the timeline: 1900, 1950 and 2000. Then we see other years below the timeline with significant facts about chimpanzees. How would we determine the number of years between the time Jane Goodall began studying chimpanzees in Tanzania and the US Fish and Wildlife Service designated chimpanzees as endangered? (We would subtract the years; we would create an equation 1988 – 1960.) Yes, we would subtract and get the answer of 28 years. We will do more problem solving as we read the book.

- ◆ As we learn more about ways to measure and represent time, we *can apply definitions and state the meaning of symbols when we communicate our answer.* You have heard people describe the time using a.m. and p.m. but do you know what those terms mean? (They mean night or day; a.m. means the morning and p.m. means the night.) Let's look at page 10. We know a day is 24 hours so 12 of those hours are a.m., which stands for *ante meridiem* and means "before noon." The other 12 hours are p.m., which stands for *post meridiem* and means "after noon." So the hours from midnight to noon are a.m. and the hours from noon to midnight are p.m., which is shown on the timeline of a day here.

◆ We can see in the photos and as we read the story that Jiggs was born at a low birth weight for a chimpanzee, which may be why his mother ignored him. She may have thought he would not live and therefore did not want to get attached to him. But the workers in the zoo try to save all animals, especially if they are an endangered species. We can see a bar graph with his weight on the vertical axis and his age in weeks on the horizontal axis. As developing problem solvers, we *can use precise language to explain how we solved a problem.* In order to practice this skill, I want you to solve this problem and I want you to be able to use precise language to explain how you found your answer:

> How much did he weigh at birth, to the nearest quarter pound? How much did he weigh at 8 weeks, to the nearest quarter pound? How much weight did he gain?

◆ In your explanation consider using language such as weight, age, weeks, pounds, quarter, half, three-quarters and other terms you could use to describe how you answered each question in the problem and how you know which operation to use. (Students should share their explanation with a peer or with the whole class.)

◆ Yes, we can see from the graph that Jiggs weighed 2 3/4 pounds at birth and he weighed 4 1/2 pounds at 8 weeks of age. We are going to subtract the weight at birth from the weight at 8 weeks and we get 1 3/4 pound.

◆ On page 16, we see a daily chart of Jiggs' activities with an example from June 29 when Jiggs had been in the care of the zoo for 124 days. Here is another problem for you to solve and practice using *precise language to explain how you solved a problem.* In order to practice this skill, I want you to solve this problem and I want you to be able to use precise language to explain how you found your answer:

> At 3:30 a.m. Jiggs drank 1 ounce of milk and went back to sleep. At 9:00 a.m. he drank 2 ounces of milk and crawled around. How many hours passed between these two events?

◆ In your explanation consider using language such as time, hours, minutes and other terms you could use to describe how you solved the problem and how you know which operation to use. (Students should share their explanation with a peer or with the whole class.)

◆ Yes, we can see from the timeline that the first event occurred at 3:30 a.m. and the second event occurred at 9:00 a.m. We can subtract the two times or we can count up from the earlier time to get 5 1/2 hours. If we subtract time we have to rewrite 9:00 as 8:60 so we can do the subtraction, much like we do when we are subtracting tens and ones. We have to subtract minutes from minutes and hours from hours. If we count up, we can count up half an hour to get to 4:00 a.m. then count up 5 hours to 9:00 a.m.

◆ On page 18, we can see another daily chart of Jiggs' activities with an example from July 28 when Jiggs had been in the care of the zoo for 153 days. Here is another problem for you to solve and practice using *precise language to explain how you solved a problem*. In order to practice this skill, I want you to solve this problem and I want you to be able to use precise language to explain how you found your answer:

> At 11:30 a.m. Jiggs drank 3 ounces of milk. At 2:00 p.m. he fell asleep without drinking his bottle. How much time passed between these two events?

◆ In your explanation consider using language such as time, hours, minutes, a.m., p.m. and other terms you could use to describe how you solved the problem. (Students should share their explanation with a peer or with the whole class.)

◆ Yes, we can see from the timeline on page 18 the first event occurred at 11:30 a.m. and the second event occurred at 2:00 p.m. We have to be careful when we are moving from a.m. to p.m. or p.m. to a.m. We can count up from the earlier time to noon to get a half hour then count up 2 more hours to 2:00 p.m. The answer would be 2 1/2 hours. We could also create a number line to show how we can move from a.m. to p.m. (See Figure 6.1.)

◆ On page 20, we can see another daily chart of Jiggs' activities with an example from September 18 when Jiggs had been in the care of

Figure 6.1

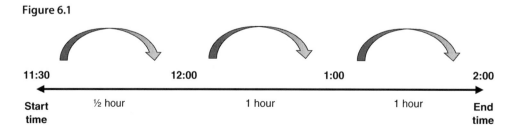

11:30	12:00	1:00	2:00
Start time	½ hour	1 hour	1 hour	**End time**

the zoo for 205 days. On this chart, Cindy used a 24-hour clock to write the time. So instead of writing 3:10 p.m. she wrote 15:10. We will use the p.m. time in our word problem though. Here is the problem for you to solve and practice *correctly applying labels and specifying units in your answer*, as well as using *precise language to explain how you solved a problem*. In order to practice these skills, I want you to solve this problem using a number line and I want you to be able to use precise language to explain how you found your answer:

> At 7:05 a.m. Jiggs watched TV and ate a banana. At 3:10 p.m. he drank milk and ate carrots. How many hours passed between these two events?

◆ In your explanation consider using language such as start time, end time, hours, minutes, a.m., p.m. and other terms you could use to describe how you solved the problem. Your number line should be labeled with the start and end times and the times you will use for your intervals as well as the arrows and the minutes or hours for each arrow. (Students should share their explanation and number line with a peer or with the whole class.)

Your students can display their number lines and compare them with their peers, as some students may have used different intervals such as moving from 7:05 to 8:05 through to 3:05 then to 3:10; others may have moved from 7:05 to 8:00 through 3:00 then to 3:10. Use other events in Jiggs' daily activities to create more word problems to provide additional opportunities to practice addition and subtraction of time intervals as well as using the practice standards.

Hershey's Milk Chocolate Weights and Measures by Jerry Pallotta

In the book *Hershey's Milk Chocolate Weights and Measures* (2002), readers are introduced to concepts of weights and measures using Hershey's products for comparisons. The illustrations use standard and nonstandard units of measurement, such as both a centimeter ruler and Reese's Pieces to represent a centimeter. Your students will recognize familiar objects, such as a gallon of chocolate milk and a 1 pound bag of Twizzlers, in order to learn the relative size of the measurement units.

> **Measurement and Data 4.MD**
>
> **Solve problems involving measurement and conversion of measurements from a larger unit to a smaller unit.**
> Know relative sizes of measurement units within one system of units, including km, m, cm; kg, g; lb, oz; l, ml; hr, min, sec. Within a single system of measurement, express measurements in a larger unit in terms of a smaller unit. Record measurement equivalents in a two-column table.

As you read the book, engage students in discussion and activities by asking the following questions based on the *I Can* statements:

◆ At the beginning of our book, we see familiar terms such as foot, inches, yard and mile, with pictures of an inch ruler and Hershey's miniature candy bars next to the ruler. We can practice *using precise language to explain how we solved a problem.* Did you know each little candy bar is 1 inch wide? How many candy bars, if the width is 1 inch, would we use to demonstrate the length of a yard and how do you know? (We would need 36 candy bars; there are 36 inches in a yard.) How many candy bars would we need to demonstrate the length of a mile and how do you know? (We would need 63,360 candy bars; there are 36 inches in a yard and 1,760 yards in a mile; we multiply 36 by 1,760 to get 63,360.) Yes, we multiply when we move from a larger unit to a smaller unit.

◆ We can also see metric length measurements in this book, such as centimeter, decimeter, meter and kilometer. Here is a centimeter ruler and Reese's Pieces candies next to the ruler. Let's practice *using precise language to explain how we solved a problem.* Did you know each Reese's Pieces candy is 1 centimeter wide? How many candies would we use to demonstrate the length of a meter and how do you know? (We would need 100 candies; there are 100 centimeters in a meter.) How many candies would we need to demonstrate the length of a kilometer and how do you know? (We would need 100,000 candies; there are 100 centimeters in a meter and 1,000 meters in a kilometer; we multiply 100 by 1,000 to get 100,000.)

◆ We can also see measures of weight, such as ounce, pound and ton. It would take 8 pieces of a Hershey's Milk Chocolate bar to weigh 1 ounce and of course a 1 pound bag of Twizzlers is 1 pound in weight. There are 16 ounces in 1 pound and 2,000 pounds in 1 ton.

How many ounces would be in 1 ton and how do you know? Do not forget to *use precise language to explain how you solved the problem.* (There would be 32,000 ounces in 1 ton; there are 2,000 pounds in a ton; there are 16 ounces in 1 pound; multiply 2,000 by 16 to get 32,000.)

◆ There are also metric weight measurements, such as milligram, gram, kilogram and metric ton. An average-sized almond is about 1 gram in weight, like the almonds in a Hershey's Milk Chocolate with Almonds candy bar. How many almonds, each weighing exactly 1 gram, would be equivalent to a metric ton and how do you know? (There would be 1,000,000 grams in a metric ton; there are 1,000 grams in a kilogram; there are 1,000 kilograms in a metric ton; we multiply 1,000 by 1,000 and get 1,000,000.)

◆ I am going to give you some measurement equivalents and we will put them into a table along with the previous equivalents we discussed. (See Table 6.1.) I want you to work with a partner and fill in the missing data. Use the information in the book, in the table and what you know about converting from a larger unit to a smaller unit to find out how many ounces in 1 gallon and how

Table 6.1

Larger Unit	Smaller Unit
1 mile	1,760 yards 5,280 feet 63,360 inches
1 kilometer	1,000 meters 100,000 centimeters
1 ton	2,000 pounds 32,000 ounces
1 kilogram	1,000 grams 1,000,000 milligrams
1 metric ton	1,000 kilograms 1,000,000 grams
1 gallon	4 quarts 8 pints 16 cups ? ounces
1 liter	100 centiliters ? milliliters

many milliliters in 1 liter. You will have to explain how you arrived at your answer showing all calculations on your paper. You should apply the problem solving skills of *applying definitions when you communicate your answer, correctly specifying units* and *using precise language to explain how you solved the problem.*

Sir Cumference and the Viking's Map by Cindy Neuschwander

In the book *Sir Cumference and the Viking's Map* (2012), Radius and his cousin Per find an ancient document with a treasure map left behind by a Viking warrior, Xaxon Yellowbearyd. One side of the document provides a clue and the other side is a map with grid lines and numbers superimposed on the map of the countryside. Your students will be introduced to coordinate axes and learn how to locate the origin and specified coordinates in order to solve a problem.

Geometry 5.G

Graph points on the coordinate plane to solve real-world and mathematical problems.

Use a pair of perpendicular number lines, called axes, to define a coordinate system, with the intersection of lines (the origin) arranged to coincide with the 0 on each line and a given point in the plane located by using an ordered pair of numbers, called its coordinates. Understand that the first number indicates how far to travel from the origin in the direction of one axis, and the second number indicates how far to travel in the direction of the second axis, with the convention that the names of the two axes and the coordinates correspond.

As you read the book, engage students in discussion and activities by asking the following questions based on the *I Can* statements:

◆ Radius and Per are lost in Angleland without a map to guide their way. On page 4, we see them at the top of a hill looking at the landscape below. They notice the land seems to be divided into four sections divided horizontally by a road and vertically by a river. They seek shelter inside a cottage built into the hillside and find a map inside a barrel. They notice the back of the map reveals a treasure belonging to Xaxon Yellowbearyd the Viking is located at (3,0) and is signed *XY*.

◆ On page 11, we can see the map with a compass rose at the top and grid lines covering the entire map. There are negative and positive numbers along a horizontal line drawn on a Viking's ax with a letter *X* on it. There are also negative and positive numbers along a vertical line also drawn on a Viking's ax with a letter *Y* on it in the center of the map. We can assume those are the two axes that Xaxon Yellowbearyd used. What does this type of map resemble? (It looks like a coordinate grid with the *x* and *y* axes.)

◆ Yes, Radius and Per are not familiar with a coordinate system but they will learn how to use it to find the treasure. According to the writing on the back of the map, the search for the treasure should start at (3,0). What do you think that means? (They should look on the map and find the coordinates 3,0 and go to that location first.) Where should they look for the 0? (They would find it at the origin, where the horizontal and vertical lines intersect.)

◆ On page 12, we see a close-up of the map and notice the circle around the cottage is actually the 0 indicating the origin. But they do not know which 3 to use for the other coordinate, since they are not familiar with how to locate coordinates on a coordinate plane. We are learning how to *apply definitions and state the meaning of symbols when we communicate our answer.* I would like you to write an explanation for Radius and Per so they would know how to read (3,0) on the coordinate plane.

◆ I like how your explanations included the definitions of the coordinates, coordinate plane, *x* and *y* axes and the meaning of the symbols. In the story they use trial and error to figure out which 3 they should use for the location. They try the 3 on Yellowbearyd's *Y* ax first but they do not find anything there. When they go back to the house, which represents the 0 in the coordinate, they follow the mile markers until they reach the mile marker corresponding with the 3 on Yellowbearyd's *X* ax and they find another clue. Since you *can use precise language to explain how you solved the problem,* we know the first coordinate represents the horizontal *x* axis and the second coordinate represents the vertical *y* axis.

◆ We see on page 13 the second clue is another coordinate, (2,–1), and is also signed *XY*, indicating the clue was left by Xaxon Yellowbearyd. They figured out the first number is found on Yellowbearyd's *X* ax so they have to find mile marker 2 near the cottage. They also realized the second number tells them which way to go on Yellowbearyd's *Y* ax so once they find mile marker 2 they have to travel 1 mile south.

◆ On page 19, we see they found another clue in the form of a coordinate, (–3,–3). As advanced problem solvers we *can correctly apply labels and specify units in our answer.* I want you and your partner to use inch grid paper and a ruler to draw a coordinate plane resembling the one in the story, with the x and y axes each extending from –6 to 6. Then write out an explanation for Radius and Per explaining how they would travel from their current location of (2,–1) to their new location at (–3,–3). Be sure to *use precise language to explain how you solved the problem.*

Continue reading the story so your students can see how Radius and Per find the treasure containing maps of Angleland. The last page of the book provides information about the coordinate system developed by Rene Descartes. Your students can explore and create maps of their community based on the Cartesian coordinate system in order to practice graphing points on a coordinate plane to solve real-world and mathematical problems.

Wrapping It Up

Students should be as diligent about learning the meaning of their math terms as they are about learning to spell words for a spelling test or studying vocabulary words in science. Provide opportunities for students to read, define, record, label and practice using precise mathematical terms. When they are ready, they will be able to understand how the terms are being used in word problems and will be able to use the terms in the explanation of their problem solving process.

7

Look for Structure

SMP 7—Look for and Make Use of Structure

Once students in primary grades learn how to recognize patterns they begin to point out patterns in their environment, including the classroom and their home. However, these are typically visual patterns, such as a sequence of red and blue cubes or shapes on a curtain. Teachers have to guide their students toward recognition of patterns and structure in other aspects of mathematics: digits end in 5 then 0 when skip-counting by fives; the number of sides of a polygon increase when the number of angles increase. As students begin to recognize this type of structure in mathematics, they will begin to use shortcuts to become more efficient at problem solving.

What Does This Standard Mean for Grades K–2 Problem Solvers?

Students who are successful at SMP 7 search for patterns that can be used to reach a solution. For example, they know how to apply counting principles of even and odd numbers based on patterns. They look for common attributes such as addends that are doubles or various shapes with four sides. They can also look for ways to group objects that are alike or apply the concept of grouping to place value when adding hundreds to hundreds, tens to tens, and ones to ones.

In order to fully apply SMP 7 when approaching a word problem, students should be able to take ownership of their procedures by using the following *I Can* statements:

- ◆ *I can find a pattern in a problem.*
- ◆ *I can look for ways to make groups in a problem.*

The Button Box by Margarette S. Reid

In the book *The Button Box* (1990), a young boy plays with the buttons that are in a box at his grandmother's house. He sorts them in different ways, such as by size, color and type. Your students can learn what it means to sort objects as well as how to use various types of attributes to classify objects into categories. Once they can identify an attribute and sort objects into categories, they can practice counting the number of objects in each category. Have a collection of buttons similar to the type of buttons in the book. You can also have buttons (real buttons or circles cut out to look like buttons) for each student that are all the same size but in three different colors. Make sure each student has the same number of buttons of each color so you can assess them.

Measurement and Data K.MD

Classify objects and count the number of objects in each category.

Classify objects into given categories; count the numbers of objects in each category and sort the categories by count.

As you read the book, engage students in discussion and activities by asking the following questions based on the *I Can* statements:

- ◆ Let's look at the picture of the box of buttons that the boy in the story just opened up at his grandmother's house. There are many buttons and they are all different. I have a collection of buttons that I brought from home. As I read the story, we will stop and look at the buttons in the book and the buttons in my collection.
- ◆ The story states that he likes to *sort* the buttons because there are so many different kinds of buttons. He will sort them into different groups, called *categories*. As I read I want you to listen for the first two different categories of buttons. (There are sparkly buttons and buttons covered with cloth.)
- ◆ Let's look at the buttons in my collection. I am going to put some of them on the document camera so we can all see them. Are any of my buttons sparkly? (Yes.) How many are sparkly? (3) Are any of my buttons covered with cloth? (Yes.) How many are covered with cloth? (2)

◆ Now I am going to read the next two pages. Listen again for some categories for sorting our buttons. (There are metal buttons and leather buttons and small buttons.) I am going to put another set of my buttons on the document camera. Do I have any metal buttons? (Yes.) How many metal buttons are in the book? You count them while I touch each one in the illustration. (1, 2, 3, 4.) How many leather buttons are in the book? You count them while I touch each one in the illustration. (1, 2, 3, 4.) There are 4 metal buttons and there are 4 leather buttons. There is the same number of metal buttons as leather buttons.

◆ Listen for the next category of buttons. What do these buttons look like? (They have designs on them. Some are silver and some are yellow or gold. Some have a flag or a star.) These are the shiny buttons that come from uniforms. I am going to put some more of my buttons on the document camera. Do I have any shiny buttons that look like the buttons in the book? (Yes.) How many? (4)

◆ Let's look at the page where the boy arranged the buttons in rows by color and by size. I am going to give you each a set of circles that we are going to use as buttons. You are going to use them to solve some word problems. Here is the first word problem:

> I have a bag of buttons. They are all the same size and shape. How can I sort them?

◆ Pour out your buttons onto your desk. What do you notice about the buttons? (They are all the same size but are different colors.) To help us solve this we *can look for ways to make groups in a problem.* So how can we sort them into groups or categories? (We can sort them by color.) We can make a category for each color. I also have a set of buttons like yours so I will put them on the document camera. Let's do one together. What is one of the colors of buttons in your set? (Red.) Okay, let's sort the red buttons first. We have a category called Red Buttons and I am going to create a chart so we can keep track of the categories. (See Table 7.1.) What other colors do you

Table 7.1

Category	Picture of Buttons	Number of Buttons
Red Buttons		
Blue Buttons		
Green Buttons		

have in your set? (Blue.) We also have blue buttons so we also have a category called Blue Buttons that I will put on our table. Are there any other colors in your set? (Green.) Yes, there are green buttons too so our last category is called Green Buttons.

◆ Now you can sort your buttons into the three different categories. I will give you three squares of paper for sorting. Sort the red buttons by putting all of the red buttons on one of the pieces of paper. How many red buttons do you have? (5)

◆ Now sort the blue buttons by putting all of the blue buttons on another piece of paper. How many blue buttons do you have? (4) Now sort the green buttons by putting them on the last piece of paper. How many green buttons do you have? (6)

◆ Now let's draw the amount of buttons in each category on our chart. (See Table 7.2.) How many buttons will I draw for the category Red Buttons? (5) Why should I draw 5 there? (Because there are 5 red buttons.) How many buttons will I draw for the category Blue Buttons? (4) Why should I draw 4 there? (Because there are 4 blue buttons.) How many buttons will I draw for the category Green Buttons? (6) Why should draw 6 there? (Because there are 6 green buttons.)

◆ Now we can use our chart to help us solve another word problem. Here is the next word problem:

I have a bag of buttons. I sorted the buttons into three categories. Are there more red, blue or green buttons?

◆ How can we fill in the rest of our chart to help us solve this problem? (We can count up the number of buttons in each category and write it in the last column.) (See Table 7.3.) What should I write for the number of red buttons? (5) What should I write for the number of blue buttons? (4) What should I write for the number of green buttons? (6)

Table 7.2

Category	Picture of Buttons	Number of Buttons
Red Buttons	⬤⬤⬤⬤⬤	
Blue Buttons	⬤⬤⬤⬤	
Green Buttons	⬤⬤⬤⬤⬤⬤	

Table 7.3

Category	Picture of Buttons	Number of Buttons
Red Buttons	●●●●●	5
Blue Buttons	●●●●	4
Green Buttons	●●●●●●	6

◆ We still have to solve the word problem, which asks if there are more red, blue or green buttons. What is your answer and how do you know? (There are more green buttons because 6 is the biggest number; because the row of green buttons is the longest; because there only 5 red buttons and 4 blue buttons, etc.)

Continue reading the story and talk about other categories of buttons in the book. You can give students buttons that vary in size but are all the same color, so they have to figure out the new category, practicing with only one attribute at a time such as size, color or type of button. Create word problems based on the new category and encourage students to *look for ways to make groups in a problem.*

The Greedy Triangle by Marilyn Burns

Use the book *The Greedy Triangle* (1994) to teach or reinforce polygons, sides and angles and to explore the concept of greed. The triangle is shown in its many important roles, such as supporting bridges, acting as a sail or a roof and being slices of pie. But the triangle becomes bored and wants to change by having one more side and angle. The Shapeshifter grants the triangle's wish and turns it into a quadrilateral. Now the triangle is shown as a computer screen, picture frames and a baseball diamond. As your students might predict, the quadrilateral is bored again and goes back to the Shapeshifter for more angles and sides. Teachers can use this book to teach defining and non-defining attributes of shapes as well as provide opportunities for students to build shapes based on particular attributes. Have toothpicks available for students to use to build the shapes in the story.

Geometry 1.G

Reason with shapes and their attributes.

Distinguish between defining attributes (e.g., triangles are closed and three-sided) versus non-defining attributes (e.g., color, orientation, overall size); build and draw shapes to possess defining attributes.

As you read the book, engage students in discussion and activities by asking the following questions based on the *I Can* statements:

◆ Have you ever heard of the word *greedy*? How would you describe someone who is greedy? (They don't share; they want everything for themselves; they are not happy with what they have; they want what someone else has.) We are going to read a book about a greedy triangle. You will each receive some toothpicks. I want you to make a triangle with your toothpicks. How many sides does a triangle have? (3) Each toothpick will represent a side so use only three toothpicks to make a triangle on your desk.

◆ I will also make a triangle on the document camera. Let's look at our triangle and see if we can think of some things shaped like a triangle. (The roof of a house, a piece of cake or pie or pizza, a sail for a sailboat, etc.) On the first page, we can see there are lots of things the triangle likes to do. We named some of these and there are even more!

◆ Now we see the triangle is feeling "dissatisfied." What do you think this word means? Look at the illustration of the triangle to help you. (He is sad. He does not feel good.) He is unhappy and goes to the Shapeshifter and asks him for two things he thinks will make his life more interesting. He asks for one more side and one more angle. I want you to use one toothpick and add a side to your three-sided figure. Now how many sides do you have? (4) When you add a side to a shape, you also add an *angle*. An angle is like the corner of a shape.

◆ The special name for any four-sided shape is a *quadrilateral*. So let's look around the classroom right now and see if we can find shapes with four sides. (There are windows, tables, desks, doors, computer screens, books, etc.) In the book, we can see the quadrilateral is happy being the square in a checkerboard, a rectangular movie screen, a computer screen, and many more things. But it is still the greedy triangle after all, so what do you predict it will do? (It will not be happy and will go to the Shapeshifter again.)

◆ Before we see what the quadrilateral is going to do, I am going to have you solve a word problem based on this story. If you *can find a pattern in a problem* then you can solve it:

> The greedy triangle had 3 sides and 3 angles. Then it went to the Shapeshifter and became a quadrilateral. The quadrilateral has 4 sides and 4 angles. Now it wants to go to the Shapeshifter again to become a pentagon. How many sides and how many angles do you think it will get? Why do you think so?

◆ If we keep track of the number of sides and angles of the shapes in the book so far, it can help us *find a pattern in the problem.* I will make a table and you help me fill in the sections. (See Table 7.4.)

◆ First let's see if we can find a relationship between the number of sides and the number of angles from the table. (There are 3 sides and 3 angles. Then there are 4 sides and 4 angles.) So the number of sides is the same as the number of angles. We can look in the rows on the table and see that relationship. Now let's see if we can find a pattern in the table in the columns that can help us solve the problem. (There were 3 sides and angles, and then there were 4. The numbers are going up by ones.)

◆ So there is a pattern in the shapes that appear in the story. If there were 3 and then 4 sides and angles, how many do you think the quadrilateral will get from the Shapeshifter when it becomes a pentagon? (5) Why do you think it will get 5 sides and 5 angles? (Because 5 comes after 4 and it is going in order. And if it will get 5 sides then it has to get 5 angles because those are always the same.)

◆ Now I'll read the next page and we'll see if we are correct. Yes, it does get one more side and one more angle so it now has 5 sides and 5 angles. I also want you to use another toothpick so you can change your quadrilateral into a pentagon.

◆ I think you are ready for another word problem based on this story. If you *can find a pattern in a problem* then you can solve it:

> The quadrilateral went to the Shapeshifter to become a pentagon. Now it is dissatisfied again and wants to become a hexagon. How many sides and how many angles do you think it will get? Why do you think so?

◆ If we continue to keep track of the number of sides and angles of the shapes in the book, it can help us *find a pattern in the problem.* Help me fill in the remaining sections. (See Table 7.5.)

Table 7.4

Shape	Number of Sides	Number of Angles
Triangle	3	3
Quadrilateral	4	4
Pentagon		

Table 7.5

Shape	Number of Sides	Number of Angles
Triangle	3	3
Quadrilateral	4	4
Pentagon	5	5
Hexagon		

- ◆ If we continue to look for a pattern in the columns, we can figure out the answer to the problem. If there were 3 and then 4 and then 5 sides and angles, how many do you think the pentagon will get from the Shapeshifter when it becomes a hexagon? (6) Why do you think it will get 6 sides and 6 angles? (Because 6 comes after 5 and it is going in order. And if it will get 6 sides then it has to get 6 angles because those are always the same.)
- ◆ Now I'll read the next page and we'll see if we are correct. Yes, it does get one more side and one more angle so it now has 6 sides and 6 angles. I also want you to use another toothpick so you can change your pentagon into a hexagon. We can also stop right now to think about how adding one more side also creates one more angle. Each time the triangle changed its shape, I asked you to add one more toothpick to make another side. But did you have to do anything to make another angle or did your shape automatically get another angle? (When we add another toothpick for a side, it makes another angle.) Yes, another angle is formed when another set of sides are joined together.
- ◆ You might think the hexagon is finally happy but the book says, "The shape became restless, dissatisfied, and unhappy with its life." What do you think it will do now? (It will go to the Shapeshifter again; it will get more sides and shapes; it will keep changing, etc.) You can see on this page that it is changing into more shapes, such as a heptagon, an octagon, a nonagon and a decagon. If we fill in our chart, we can see how many sides and angles are in those other shapes. (See Table 7.6.)

Have your students use their toothpicks to create the heptagon and the rest of the shapes, noticing how another angle is formed as they add another side. Or you can create word problems based on those for the pentagon and hexagon so your students can practice explaining how they can use a pattern to solve the problem. As a follow-up activity, give students several cut-outs of

Table 7.6

Shape	Number of Sides	Number of Angles
Triangle	3	3
Quadrilateral	4	4
Pentagon	5	5
Hexagon	6	6
Heptagon	7	7
Octagon	8	8
Nonagon	9	9
Decagon	10	10

each shape in the story as well as circles, ovals and figures that are not closed, varying in size and color. Include some polygons that are regular and some that are not so they realize sides are not always congruent. Then students can sort the shapes according to attributes, such as number of sides, number of angles, color, closed figures and shapes that do not have straight sides.

Earth Day—Hooray! by Stuart J. Murphy

In the book *Earth Day—Hooray!* (2004), a group of children are cleaning up a neighborhood park for an Earth Day activity. They realize if they take the cans to the recycling center, they can earn money to buy flowers to plant in the park. At school their club adviser helps them keep track of their cans in groups of tens, hundreds and eventually thousands in order to achieve their goal of turning in 5,000 cans for money. Your students can look for ways to make groups as they use models, drawings and strategies based on place value to solve problems.

Number and Operations in Base Ten 2.NBT

Use place value understanding and properties of operations to add and subtract.

Add and subtract within 1,000, using concrete models or drawings and strategies based on place value, properties of operations and/or the relationship between addition and subtraction; relate the strategy to a written method. Understand that in adding or subtracting three-digit numbers, one adds or subtracts hundreds and hundreds, tens and tens, and ones and ones; and sometimes it is necessary to compose or decompose tens or hundreds.

As you read the book, engage students in discussion and activities by asking the following questions based on the *I Can* statements:

- ◆ The children in this story are part of the Maple Street School Save-the-Planet Club and they are cleaning up a park near their school. What kind of garbage do you see the children picking up on pages 4 and 5? (They are picking up wrappers, scraps of paper, newspapers, bottles, cans and cups.) On page 6, Ryan realizes that the park would look better with some flowers at the entrance. How could they get money to buy flowers? (They could ask their parents, use their allowance, do a fundraiser at school, etc.) Have you heard of a recycling center? Ryan talks about taking the aluminum cans they are collecting to a recycling center so they can get money for the cans. These centers buy cans from people and then use them to make new cans. What do you think Ryan and his friends would do with the money they get from turning in the cans? (They could buy the flowers to plant in the park.)
- ◆ They tell their club adviser about it and she said they should collect 5,000 cans! How could they keep track of that many cans? I am going to give you a minute to talk to a friend about how the children can keep track of all of the cans they collect so they will know when they reach 5,000. (They can count them, put them in stacks, write down how many they collect, etc.)
- ◆ Thanks for sharing all of the ways you discussed with your friend. I am going to show you page 10 so you can see how they sorted the cans and put them into groups of ten. Why do you think it will be easier for them to count the cans if they are in groups of ten? (Because they can count by tens faster than by ones.) We know good problem solvers *can look for ways to make groups in a problem*. It is easier and faster to count by tens and then they can keep track of the groups of hundreds as they collect more cans. How many groups of ten will they need in order to have a group of one hundred cans? (They would need ten groups of ten.)
- ◆ As I read the book, I am going to tell you the groups of cans they are collecting, in ones, tens, hundreds and eventually in thousands, but you will have to keep track of the amounts. Let's think about how we *can look for ways to make groups in a problem*. What tools can you use to keep track of the groups of cans? (We can draw pictures, make a graph, use a calculator, etc.)
- ◆ Let's try drawing pictures to help us solve some of these word problems. I am not going to show you the picture on page 11 until

after you have drawn a picture so you can check your work. I will write a word problem that goes with this page so you can look at it as you draw your picture:

> The children collected 3 big bags of 100, 5 small bags of 10 and 9 single cans. How many cans did they collect so far?

◆ Think about how you could draw bags and label them. Just draw a rectangle for a can so it won't take long. When you are finished, show your drawing to a friend sitting near you and figure out how many cans the children collected. Write the answer on the top of your paper and circle it.

◆ How did you know that there were 359 cans? (We added the hundreds, which was 300. Then we added the tens, which was 50. Then we added 300 + 50 + 9, which is 359.) You remembered to group all of the hundreds together, then group the tens together and then add them up with the ones. How did *making groups in a problem* help you figure out the problem? (It is easier to group the hundreds together, group the tens together and group the ones together as we add the numbers.)

◆ As I continue to read the story, we can see on page 17 the children had to start all over because the trash collector took away their cans. But now they are starting over again with 56 cans. On pages 17, 18 and 19, how are they able to get more cans since they already cleaned up the cans from the park? (They are putting up signs around the school, going to each classroom and picking up cans around the neighborhood.)

◆ I am going to give you another word problem to solve and this time you *can look for ways to make groups in a problem* by using base ten blocks for you and your partner to use as models for the hundreds, tens and ones. Here is the problem that has two parts to the solution:

> The children brought the cans they collected to school to count at recess. There were 6 bags of 100, 3 bags of 10 and 5 single cans. How many cans did they count at recess? How many cans did they collect so far if they combine those cans with the 56 cans they already collected?

◆ Now use your base ten blocks to find out how many cans they counted at recess. Don't forget to group the hundreds together, the tens together and the ones together before you add them up to find your total. How many cans did they count at recess? (They counted 635 cans.) How did *making groups in a problem* help you figure out the problem? (It is easier to group the hundreds blocks together to see there were 600, then we grouped the tens blocks together to see there were 30 and then we had 5 ones. We added $600+30+5$ together to get 635.)

◆ Now we have to solve the second part of the problem. If they counted 635 cans at recess and they already had 56 cans, how many cans have they collected altogether? You might need to compose a ten when you are adding the tens and ones. Use more base ten blocks to add to the 635 that you already have on your base ten mat.

◆ How many cans did they collect altogether? (They collected 691 cans.) How did using the base ten blocks and mat help you figure out this part of the problem? (We added 6 ones to the 5 ones that we already had on our mat. That made 11 ones so we exchanged 10 of those for a tens block. Then we put that tens block and 5 more tens blocks on the tens section with the 3 tens that we already had there. That made 9 tens. We didn't add any more hundreds so that made 691.)

◆ You are doing a great job keeping track of the hundreds, tens and ones. I think you're ready to try the thousands now! We are going to do one more word problem but we will use paper and pencil. Here is the word problem that has two parts:

> The children collected cans from students and teachers all week. They collected 1 bag of 1,000, 4 bags of 100, 8 bags of 10 and 3 single cans. How many cans did they collect that week? How many cans will they have if they add that number to the 691 cans they already collected?

◆ Let's do the first part together. How many bags of 1,000 did they collect? (1 bag.) So we are going to write 1,000 on our paper. I will write it on the Smart Board so you can see if you wrote it correctly. How many bags of 100 did they collect? (4 bags.) How will we figure out how many hundreds we'll have in our problem? (We have to multiply 100 by 4 and we get 400.)

◆ Write 400 below 1,000 on your paper, but line up the hundreds. We will do the tens now. There are 8 bags of 10. How will we figure out how many tens we'll have in our problem? (We have to multiply 10 by 8 and we get 80.)

◆ Write the 80 below the 400 but line up the tens. There were also 3 single cans so we have to write 3 below the 80 but line it up with the ones and then add the numbers:

```
1,000
 400
  80
   3
```

◆ How many cans did they collect that week? (They collected 1,483 cans.) How did using paper and pencil help you figure out the first part of the problem? (We could figure out how many hundreds and tens using mental math and then write the other numbers on the paper.) How did *making groups in a problem* help you figure out the problem? (It is easier to group the hundreds together to see there were 400, then we grouped the tens together to see there were 80 and then we had 3 ones; it made it easier to add them on our paper because we have to add the ones together, add the tens together and add the hundreds together.)

◆ We have one more part of the problem to solve. How many cans have they collected altogether? Let's use our calculator to solve this problem but I still want you to write the equation on the paper so you can see the numbers you will be putting into the calculator. What is the equation for this part of the problem? (It is $1,483 + 691 = ?$) Let's all write that equation on our paper. Now put that into your calculator. Be sure to press all of the number keys correctly for 1,483 and then the + key. You don't have to press anything for the comma that is between the 1 and the 4. Now put in 691 and press the = key. You should see 2,174 on your screen.

You can continue reading the book and challenge your students to solve the final problem in the book, which is on pages 28 and 29, where the children finish collecting the cans and get their grand total to be able to turn in the cans to the recycling center for money. Your students can work on this last two-part problem with a partner in class or for homework.

What Does This Standard Mean for Grades 3–5 Problem Solvers?

Once students learn how to recognize patterns in number sequences and in the structure of regular polygons they are ready to look for ways to use patterns and structure to become more efficient at problem solving. They may notice relationships among operations or how problems are similar but the numbers may be larger.

In order to fully apply SMP 7 when approaching a word problem, students in intermediate grades should be able to take ownership of their procedures by using the following *I Can* statements:

- ◆ *I can find a pattern or structure within a problem.*
- ◆ *I can determine the significance of information in the problem.*

Hershey's Kisses from Addition to Multiplication by Jerry Pallotta

In the book *Hershey's Kisses from Addition to Multiplication* (2004), Hershey's Kisses candies are used to represent quantities to be added and multiplied. The examples are shown with equal groups as well as with arrays so the relationship between addition and multiplication can be seen. There are also explanations of how to perform multi-digit multiplication problems using the knowledge of two one-digit products and place value.

Operations and Algebraic Thinking 3.OA

Represent and solve problems involving multiplication and division.
Use multiplication and division within 100 to solve problems in situations involving equal groups, arrays and measurement quantities, e.g., by using drawings and equations with a symbol for the unknown number to represent the problem.

As you read the book, engage students in discussion and activities, with a multiplication table for each student, by asking the following questions based on the *I Can* statements:

- ◆ If we have a group of Hershey's Kisses candies, what are some ways we could determine the quantity? (We could estimate; we could count the candies; we could put them into groups and skip-count.) We can see on the fourth page there are 15 candies. On the fifth page, the candies are rearranged into 3 groups of 5 candies. This makes it easier to count them by fives or using the addition equation, $5+5+5=15$. On the sixth page, the candies are rearranged into 5 groups of 3 candies. This would make a longer addition equation, $3+3+3+3+3=15$. But there is an easier way to represent these two examples with multiplication.

◆ As developing problem solvers we *can find a pattern or structure within a problem.* On the seventh page, there are two multiplication equations, $3 \times 5 = 15$ and $5 \times 3 = 15$. How are these two multiplication equations related to the two addition equations on the previous pages? Do you see a pattern or structure within the equations? The terms in the addition equation are addends and the terms in the multiplication equation are called factors. (The first addition equation is related to the first multiplication problem; the factor 3 represents the number of groups in the addition equation and the factor 5 represents the number of candies in each group. The second addition equation is related to the second multiplication problem; the factor 5 represents the number of groups in the addition equation and the factor 3 represents the number of candies in each group.)

◆ We can see how addition and multiplication can be related and how to represent a repeated addition equation, with equal groups, as a multiplication equation with two factors. On the eighth page, there are the 15 candies again but arranged in an array with 3 rows with 5 candies in each row. There are 5 columns and 3 candies in each column. We can look at the array and we *can find a pattern or structure within a problem.* Rows are horizontal and columns are vertical in an array.

◆ We can also look for patterns and structure in a multiplication table, which is another example of an array. Let's look for patterns in the rows and columns first. What are some patterns you notice? (The first row and first column are the numbers 1 through 10; the second row and second column are identical, so are the third row and third column, etc.; every other column has all even numbers and other columns alternate odd and even numbers.)

◆ Now we are ready to solve a problem using what we have learned about multiplication:

> There are 20 Hershey's Kisses on the table. How many different ways can you put the candies into equal groups? Show all of the groups and represent each group with a multiplication equation.

◆ You should have found at least four different ways and represented them with the following multiplication equations: $2 \times 10 = 20$, $10 \times 2 = 20$, $4 \times 5 = 20$, $5 \times 4 = 20$. Did anyone include 1 group of 20 candies, $1 \times 20 = 20$, or 20 groups of 1 candy, $20 \times 1 = 20$?

◆ Let's look at an equation with three factors on page 26. I am going to show you the equation and the solution and I want you to *determine the significance of information in the problem.*

$$2 \times 5 \times 7 = ?$$
$$(2 \times 5) \times 7 = ?$$
$$10 \times 7 = 70$$

◆ What is the significance of the parentheses in the second equation and why do you think so? (They show you what to multiply first; you multiply 2×5 to get 10 then you multiply 10×7 to get 70.) Yes, the significance of the parentheses is to indicate which factors to multiply first. We can use them in a multiplication equation with three factors just like we have used them in an addition equation with three addends.

◆ Let's look at an equation with four factors on page 29. I am going to show you the equation and the solution and I want you to *determine the significance of information in the problem.*

$$2 \times 3 \times 4 \times 5 = ?$$
$$(2 \times 3) \times (4 \times 5) = ?$$
$$6 \times 20 = 120$$

◆ What is the significance of the parentheses in the second equation and why do you think so? (They show you what to multiply first; you multiply 2×3 to get 6 and you multiply 4×5 to get 20; then you multiply 6×20 to get 120.) Yes, the significance of the parentheses is to indicate which factors to multiply first. We can use them in a multiplication equation with four or more factors just like we have used them in an addition equation with three or more addends.

◆ We *can step back and look at a problem in a new way, making adjustments as needed.* I want you to work with a partner to rewrite the equation with four factors so your parentheses are used to group different factors. It is okay if you have to rearrange the factors, the product will remain the same. If you recall from the Commutative and Associative Properties of Addition, we can rearrange addends and group them in different ways and the sum will remain the same. This works for the Commutative and Associative Properties of Multiplication as well.

Continue creating opportunities for students to solve problems with equal groups and arrays using drawings or models along with multiplication equations. Students should also be able to use parentheses to group factors in equations with more than three factors and rearrange factors in various ways.

If You Hopped Like a Frog by David M. Schwartz

In the book *If You Hopped Like a Frog* (1999), the author creates comparisons between the abilities of animals and those of a human, if the human had the same abilities as the animal. A frog can jump 20 times its body's length. If we applied that to a human, how far could the human jump in one leap? Your students will find themselves using concepts of measurement conversions as they try to figure out how the author made his comparisons.

> **Measurement and Data 4.MD**
>
> **Solve problems involving measurement and conversion of measurements from a larger unit to a smaller unit.**
> Use the four operations to solve word problems involving distances, intervals of time, liquid volumes, masses of objects and money, including problems involving simple fractions or decimals, and problems that require expressing measurements given in a larger unit in terms of a smaller unit.

As you read the book, engage students in discussion and activities, with a multiplication table for each student, by asking the following questions based on the *I Can* statements:

◆ This book is full of interesting facts comparing the feats of animals and insects to human feats. For example, on the first page it states that if you hopped like a frog you would be able to jump from home plate to first base in one leap. How would the author be able to determine the length of the leap of a human? In the back of the book, the author reveals his calculations so, as we read the book, I will give you some of his answers and you will have to figure some out on your own. We will have to *determine the significance of information in the problem* in order to find the answers.

◆ In the back of the book, I read that a frog can jump 20 times its body's length. So if the frog were 2 inches long, how far could it jump? (It could jump 40 inches.) Yes, we would just multiply 2×20

to find out how far it could jump. How did the author, David Schwartz, figure out that a boy could leap from home plate to first base in one leap? (He could figure out how tall a boy would be then multiply the height by 20; he knew the distance from home plate to first base would be about the same distance.)

◆ Yes, he used 4 1/2 feet for the average boy and then multiplied that by 20 to get 90 feet, which is the distance between home plate and first base. What information did you determine was significant for this problem? (The information about how far a frog can jump compared to its length; the height of the person.)

◆ So for each of the examples in the book we have to *determine the significance of information in the problem* in order to find the answers. It will be the information about the animal and then related information about a boy or girl in order to make the comparison. As we do a few problems you will also be able to *find a pattern or structure within the problem.*

◆ On the next page, we find out that if you were as strong as an ant you could lift a car. What is the significant information the author needed to be able to make this comparison? (He needed to know how much an ant could lift compared to its size; he needed the size of an average girl like the one in the illustration; he needed to find something the same weight as the amount the girl could lift if she could lift objects like an ant.)

◆ We will also use some measurement conversions in our problems. The back of the book reveals that an ant could lift 50 times its own weight. The weight he used for a girl is 60 pounds. Let's figure out how many pounds and tons she could lift. What operation did we use to figure out the distance the boy could leap once we knew his height and the amount a frog could jump based on its body length? (We multiplied the numbers.)

◆ So there is a pattern or structure to these types of problems. Let's use multiplication with the weight of the girl and the amount an ant could lift based on its body weight. (We would multiply 50×60 and we would get 3,000 pounds.) Now we have to figure out how many tons are equivalent to 3,000 pounds. When we convert from a smaller unit to a larger unit we divide. We know 1 ton is equivalent to 2,000 pounds. So if we divide 3,000 by 2,000 we get 1 1/2 tons. The illustration shows a girl lifting a car that weighs about 1 1/2 tons!

◆ Now we have a comparison with a dinosaur. If your brain were the size of a brachiosaurus, it would be smaller than the size of a pea.

We first have to *determine the significance of information in the problem.*
What information do we need to know? (We need to know the
weight of a brachiosaurus and the weight of his brain; we need to
know the weight of an average boy.) This time we have a very large
object, a brachiosaurus, and its brain, which is obviously smaller
than the dinosaur's body, for the first comparison. These
measurements are given in metric units. The dinosaur weighed
80,000 kilograms and its brain weighed 200 grams. How could we
determine the relationship between those two weights so we could
use it to compare to a human's brain? (We have to figure out how
many times bigger the dinosaur is than its brain.)

◆ Yes, but first we have to convert the units so they are the same. We
know the prefix kilo means thousand so we know there are 1,000
grams in a kilogram. We can use that to divide 200 by 1,000 and we
determine there are 0.2 kilograms in 200 grams. Now we can use
the amounts of 80,000 kilograms and 0.2 kilograms. In the previous
problems we were learning about animals that can perform feats
that were greater than their body length or weight so we multiplied
our numbers. Now we have an example of a large animal with a
small brain. The pattern for this type of problems is to divide the
amounts. What would be our answer? (We would divide 80,000 by
0.2 and we get 400,000.)

◆ The weight of the boy for the comparison is 30 kilograms. Again,
we want to use the same unit so let's convert 30 kilograms to
grams. When we convert from a larger unit to a smaller unit we
multiply. We know 1 kilogram is equivalent to 1,000 grams. So if
we multiply 30 by 1,000 we get 30,000 grams. Now we divide that
number by 400,000 to get 0.075 grams. That is less than the weight
of a pea, which is what was used for the comparison.

◆ Now you can do one by yourself but it will not have as many
conversions as the one we just completed. If you swallowed like a
snake you could eat a hot dog thicker than a telephone pole. I will
give you some questions and prompts for you to use as you work
through this problem. First, *determine the significance of information in
the problem.* What information do we need to know about snakes
and about the boy? (We need to know the size of the snake's head
and the size of the food it can eat; we need to know the size of a
boy's head.)

◆ Second, if we know the snake's head is 1 inch wide and it can
swallow a gopher 2 inches wide, and we know the boy's head is
5 inches wide, how can we use this information to *find a pattern or*

structure within the problem? (We first find the relationship between what a snake can swallow based on the size of its head and then use that to figure out the size of an object a boy could swallow.)

◆ Now you are ready to use the information and the pattern to solve the problem: What could the boy swallow if he swallowed like a snake?

There are many more examples in the book you could use with your students to create word problems involving comparisons with the way an animal jumps, eats, flicks its tongue, cranes its neck and more. There are opportunities for your students to convert measures from larger to smaller units and to find patterns within the procedures for determining the comparisons.

Cheetah Math by Ann Whitehead Nagda

In this book, *Cheetah Math: Learning about Division from Baby Cheetahs* (2007), your students will love seeing how workers in the San Diego Zoo cared for two cheetah cubs when their mother became ill. The math concept shown on each page is division used when determining how much milk to feed the cubs, the number of feedings in a day, their age, weight and other essential calculations used by the caregivers at the zoo.

Number and Operations in Base Ten 5.NBT

Perform operations with multi-digit whole numbers and with decimals to hundredths.

Find whole-number quotients of whole numbers with up to four-digit dividends and two-digit divisors, using strategies based on place value, the properties of operations and/or the relationship between multiplication and division. Illustrate and explain the calculation by using equations, rectangular arrays and/or area models.

As you read the book, engage students in discussion and activities, with a multiplication table for each student, by asking the following questions based on the *I Can* statements:

◆ We are going to read about two cheetah cubs, Majani and Kubali, raised at the San Diego Zoo by zookeepers when the mother cheetah became ill and could not raise them. We will also explore concepts of division as we learn how the cubs were cared for on a daily basis. Let's review a few terms as we see how the nursery keeper, Janet, determines how to feed the cubs. She mixed powdered kitten formula with water to make 18 ounces of milk to feed the two cubs. She had to divide by 2 in order to determine each cub should receive 9 ounces of milk. The dividend, the number being divided into equal parts, in this problem is 18. The divisor, the number of parts, in this problem is 2. The quotient, the result of the division, in this problem is 9.

◆ We can also think about the relationship between division and multiplication in order to find the quotient. On page 10, we see each cub drinks 15 ounces of milk a day, divided into 5 feedings. How much milk does a cub receive at one feeding? Which number is the dividend and how do you know? (The dividend is 15 because that is the number being divided into equal parts.) Which number is the divisor and how do you know? (The divisor is 5 because that is the number of equal parts.) What multiplication fact could we use to find the quotient and why? (We could use $5 \times 3 = 15$ or $3 \times 5 = 15$; division and multiplication are related; we can think of the way 3, 5 and 15 are related.)

◆ Yes, we can use the multiplication facts for 3 and 5 to help us find the quotient of 3. The cubs will get 3 ounces at each feeding. On page 12, we see Janet giving them some meat now that they are 49 days old. How many weeks old are the cubs? We can think of perfect squares to help us find this quotient. What multiplication fact could we use to find the quotient and why? (We could use $7 \times 7 = 49$; division and multiplication are related; we know 49 is a perfect square.)

◆ Yes, we can use our knowledge of perfect squares to find the quotient of 7. The cubs are 7 weeks old. Now we will *find a pattern or structure within the problem* and use repeated subtraction to find the quotient. At the zoo they pair up dogs with the cheetahs to help keep them calm. One of the cheetahs was 11 pounds when he was introduced to the dog, which was 44 pounds. How much bigger was the dog than the cheetah? When we use repeated subtraction we can see a pattern in the way we keep subtracting the same number until we get 0. How many times did we subtract 11 from 44? (We subtracted it 4 times.) The answer is the dog was 4 times bigger than the cheetah.

◆ Sometimes when we use repeated subtraction we do not end up with 0. If that occurs we *can determine the significance of the information in the problem.* Let's look at another pairing of a dog with a cheetah. This time the dog weighed 49 pounds and the cheetah weighed 16 pounds. How much bigger was the dog than the cheetah? When we use repeated subtraction we can see the pattern again in the way we keep subtracting the same number until we get to 0 or to a number less than 16. How many times were we able to completely subtract 16 from 49 and how much was left? (We could subtract it 3 times and we had 1 left over.) What is *the significance of this information in the problem*? What does the 3 represent? (The 3 means the dog was 3 times bigger than the cheetah.) The 1 is not significant in this problem since we were able to determine the dog was 3 times bigger than the cheetah.

◆ Another strategy we could use to divide numbers is based on place value. Now that the cubs are older they only eat meat. Together the 2 cheetahs eat 44 ounces in a day. How much does each cheetah eat? We can *find a pattern or structure within the problem* based on tens and ones. First we decompose 44 into $40 + 4$. Now we can divide the tens and ones each by 2. What is 40 divided by 2? (It is 20.) What is 4 divided by 2? (It is 2.) Our answer is $20 + 2$, which is 22 in standard form. So they each eat 22 ounces of meat in a day.

◆ Another strategy is the area model for division, which allows us to *determine the significance of the information in the problem*. The cheetahs are now 105 days old. How many weeks old are they? We know there are 7 days in a week so we can think about our multiplication facts for 7. Go ahead and think of any multiplication fact for 7 that you know. (I know $7 \times 5 = 35$). Okay, let's use that one to get us started. I am going to start using the area model as you give me the numbers:

$$
\begin{array}{c}
 5 \\
7\ \boxed{\begin{smallmatrix}105\\ -35\end{smallmatrix}}\ \boxed{70}
\end{array}
$$

◆ Let's look at what we have so far to *determine the significance of the information in the problem*. We started with the dividend, 105, to represent their age in days. We have the divisor, 7, on the side to represent the number of equal groups, which is for the number of days in a week. We used the multiplication fact $7 \times 5 = 35$. When we subtract 35 from 105 we still have 70 left over. What multiplication fact for 7 could we use? (We can use $7 \times 10 = 70$.)

Yes, I will put the 70 in the box to subtract it from 70 so we have no remainder.

$$7 \quad \begin{array}{cc} 5 & 10 \\ \boxed{\begin{array}{c} 105 \\ -35 \end{array}} & \boxed{\begin{array}{c} 70 \\ -70 \end{array}} \end{array}$$

◆ Let's look at what we have so far to *determine the significance of the information in the problem.* What is the significance of the numbers on top of the boxes? (They show how many groups of 7 there are in 105; we have to add them so we get 15.) What is the answer to our problem? (The cheetahs were 15 weeks old.)

There are more division problems in the book your students can use to practice the many strategies for dividing multi-digit numbers. They should also practice looking for patterns in their calculations and explaining the significance of the information in the problem, especially when interpreting remainders.

Wrapping It Up

Encourage your students to look for patterns in their environment and seek opportunities to utilize patterns in their problem solving. Students should have opportunities to notice number patterns or observe the structure of regular polygons. You can also provide reinforcement of mathematical terms so students have a way to discuss, describe and explain their patterns and use of structure.

8

Apply Repeated Reasoning
SMP 8—Look for and Express Regularity in Repeated Reasoning

Teachers should teach shortcuts with caution. Often students rely on the shortcut but are unable to explain why the shortcut works, such as moving a decimal point to the right to multiply by 10. However, when students discover a shortcut on their own, they tend to understand the mathematics behind it. Teachers should explicitly teach students how to recognize opportunities in word problems for using a shortcut, such as when a calculation is repeated. For example, if a word problem involves adding $5+5+5+5$, the students should stop and discuss how it would be more efficient to count by fives or to multiply instead of writing out all of the addends. In order to find shortcuts, students have to pay attention to the details of the problem as well as be aware of the goals of the problem. In the previous example, it does not matter how the problem solver arrived at the answer of 20, but that they understand there are efficient ways to arrive at the same answer to satisfy the goal of the problem.

What Does This Standard Mean for Grades K–2 Problem Solvers?

Students who are successful at SMP 8 are learning efficient strategies and shortcuts to be used in word problems. They recognize the problem solving situations for addition and subtraction (see Table A1.1 in Appendix) based on the format of the problem. They pay attention to the details of the problem while they keep the goal of the solution of the problem in mind.

In order to fully apply SMP 8 when approaching a word problem, students should be able to take ownership of their procedures by using the following *I Can* statements:

- ◆ *I can look for repeated calculations.*
- ◆ *I can create a shortcut.*
- ◆ *I can pay attention to details while I think about the goal of the problem.*

Bunches of Buttons: Counting by Tens by Michael Dahl

In this book, *Bunches of Buttons: Counting by Tens* (2006), Billy collects buttons in a jar, finding them in various places in his house. As he finds the buttons, he counts them in groups of ten, eventually collecting 100 buttons. In the illustrations of the collections of buttons, each set of ten is a different color in order to distinguish each set. On each page, the decades are also shown in numeral form with ten dots above each numeral, from 10–100. Your students can practice writing, representing and counting by tens.

Counting and Cardinality K.CC

Know number names and the count sequence.

Count to 100 by ones and by tens.

Write numbers from 0 to 20. Represent a number of objects with a written numeral 0–20 (with 0 representing a count of no objects).

As you read the book, engage students in discussion and activities by asking the following questions based on the *I Can* statements:

- ◆ In this book, Billy likes to collect buttons. Let's see how many he can find in his house. On page 3, how many buttons does he have in his jar? (0) I want you to write the numeral 0 on your dry erase board. If he has *zero* buttons that means he has no buttons in his jar. Now you can erase the 0.
- ◆ On page 4 he finds 10 red buttons. I want you to write the numeral 10 on your dry erase board. Which digits did you have to write to make 10? (We had to write a 1 and a 0.)
- ◆ You can erase the 10. Let's see how many buttons he finds on page 6. How many does he find and what color are the buttons? (He finds 10 orange buttons.) He put the 10 orange buttons in his jar with the 10 red buttons. We are going to count the buttons in

the jar. Ready? 1, 2 … 20. There are now 20 buttons in his jar. Could we have counted them another way? Let's see if we *can create a shortcut* for the next page. Write 20 on your dry erase board so we can be sure everyone is writing it correctly. Which digits did you have to write to make 20? (We had to write a 2 and a 0.)

◆ You can erase the 20. Let's see how many buttons he finds on page 8. How many does he find and what color are the buttons? (He finds 10 purple buttons.) He put the 10 purple buttons in his jar with the 10 red buttons and the 10 orange buttons. We are going to count the buttons in the jar. Ready? 1, 2 … 30. There are now 30 buttons in his jar. Could we have counted them another way? I want to see if we *can create a shortcut* for this page because it is taking longer to count the buttons by ones. Is there a faster way to count the sets of buttons? (We can count them by tens.) Let's try it: 10, 20, 30. That was a great shortcut!

◆ Write 30 on your dry erase board so we can be sure everyone is writing it correctly. Which digits did you have to write to make 30? (We had to write a 3 and a 0.)

◆ You can erase the 30. Let's see how many buttons he finds on page 10. How many does he find and what color are the buttons? (He finds 10 green buttons.) He put the 10 green buttons in his jar with the 10 red buttons, 10 orange buttons and 10 purple buttons. We are going to count the buttons in the jar. Ready? 1, 2 … 40. There are now 40 buttons in his jar. How can we count them with our shortcut? (We can count them by tens.) Let's try it: 10, 20, 30, 40.

◆ Write 40 on your dry erase board so we can be sure everyone is writing it correctly. Which digits did you have to write to make 40? (We had to write a 4 and a 0.)

◆ You can erase the 40. Let's see how many buttons he finds on page 12. How many does he find and what color are the buttons? (He finds 10 pink buttons.) He put the 10 pink buttons in his jar with the 10 red buttons, 10 orange buttons, 10 purple buttons and 10 green buttons. We are going to count the buttons in the jar. Ready? 1, 2 … 50. There are now 50 buttons in his jar. How can we count them with our shortcut? (We can count them by tens.) Let's try it: 10, 20, 30, 40, 50.

◆ Write 50 on your dry erase board so we can be sure everyone is writing it correctly. Which digits did you have to write to make 50? (We had to write a 5 and a 0.)

◆ You can erase the 50. Let's see how many buttons he finds on page 14. How many does he find and what color are the buttons?

(He finds 10 blue buttons.) He put the 10 blue buttons in his jar with the 10 red buttons, 10 orange buttons, 10 purple buttons, 10 green buttons and 10 pink buttons. We are going to count the buttons in the jar. Ready? 1, 2 … 60. There are now 60 buttons in his jar. How can we count them with our shortcut? (We can count them by tens.) Let's try it: 10, 20, 30, 40, 50, 60.

◆ Write 60 on your dry erase board so we can be sure everyone is writing it correctly. Which digits did you have to write to make 60? (We had to write a 6 and a 0.)

◆ You are getting very good at this so I'm going to give you a little word problem to solve:

> Billy has 60 buttons in his jar. He finds 10 yellow buttons. Now how many buttons will he have in his jar?

◆ In order to solve this problem, think about the counting pattern. He always finds the buttons in sets of 10 so it is easy for him to count by tens. This is a type of *repeated calculation* because he adds 10 more buttons in his jar every time. Who thinks they know the answer to the problem? (He will have 70 buttons in his jar.) How did you know? (If we count by tens we will say 70 next.)

◆ You can erase the 60. Let's check pages 16 and 17 to see if we are correct. He finds 10 yellow buttons and puts them in the jar with the other 60 buttons. Let's count the buttons in the jar by tens, 10, 20 … 70. Can you write the numeral 70 on your dry erase board? Which digits did you have to write to make 70? (We had to write a 7 and a 0.)

◆ Here is another little word problem to solve but it has two parts to the solution:

> Billy has 70 buttons in his jar. How many buttons do you think he will find next? How many buttons will he have in his jar altogether?

◆ The first part of this word problem asks about the number of buttons he will find next. Think about the pattern in the book. How many buttons does he find each time? (10) So we predict he will find 10 buttons. Now we can solve the second part of the problem. How many buttons will he have if he adds the 10 buttons to the 70 buttons he already has in his jar? (80) How do you know? (If we count by tens we will say 80 next.) That's right. It is the same as adding 10 more, which is the *repeated calculation* in the book. Counting by tens to get to 80 is a *shortcut* instead of counting by ones.

◆ You can erase the 70. Let's check pages 18 and 19 to see if we are correct. He finds 10 white buttons and puts them in the jar with the other 70 buttons. Let's count the buttons in the jar by tens, 10, 20 … 80. Can you write the numeral 80 on your dry erase board? Which digits did you have to write to make 80? (We had to write an 8 and a 0.)

◆ We're almost done with the story. How many buttons do you think he will have in his jar on the next page? (90) How do you know? (If he finds 10 more buttons then he will count by tens again. If we count by tens we will say 90 next.)

◆ You can erase the 80. Let's check pages 20 and 21 to see if we are correct. He finds 10 brown buttons and puts them in the jar with the other 80 buttons. Let's count the buttons in the jar by tens, 10, 20 … 90. Can you write the numeral 90 on your dry erase board? Which digits did you have to write to make 90? (We had to write a 9 and a 0.)

◆ What if Billy keeps looking and finds another 10 buttons in his house? How many buttons would he have if he put 10 more in his jar of 90 buttons? Erase your 90 and write the answer on your dry erase board. (100) Let's see if you are correct by looking at pages 22 and 23. He finds 10 black buttons and adds them to his jar of 90 buttons. Let's count the buttons in the jar by tens, 10, 20 … 100!

Your students can practice writing all of the numerals from 0 to 100 by tens. They can count math manipulatives that are in sets of ten, such as base ten blocks and Unifix cubes, or collections of objects they brought from home. You can read the other books by the same author that are about counting by tens as well as his books about counting by fives and twos.

The King's Commissioners by Aileen Freidman

In this book, *The King's Commissioners* (1994), the King has so many royal commissioners that he constantly loses track of who is taking care of what. He decides to count them so he can begin to organize their duties. He asks his First and Second Royal Advisors to help count the commissioners as they come in the throne room. As they come in one at a time, the King tries to count them in his head while the Royal Advisors make tallies. When the King loses count and can't make sense of the tallies, his daughter asks the

Royal Organizer to put all of the commissioners in rows of ten so they will be easier to count. Your students can learn how to read and write two-digit numbers, organize quantities into tens with any leftovers counted as ones, and group numbers by twos and fives.

Number and Operations in Base Ten 1.NBT

Understand place value.

Understand that the two digits of a two-digit number represent amounts of tens and ones. Understand the following as special cases:

a. 10 can be thought of as a bundle of ten ones — called a "ten."
b. The numbers from 11 to 19 are composed of a ten and one, two, three, four, five, six, seven, eight, or nine ones.
c. The numbers 10, 20, 30, 40, 50, 60, 70, 80, 90 refer to one, two, three, four, five, six, seven, eight or nine tens (and 0 ones).

As you read the book, engage students in discussion and activities by asking the following questions based on the *I Can* statements:

◆ This story is about a King who has many, many royal commissioners who are in charge of things such as Flat Tires and Things That Go Bump in the Night. One day he decides to count them all so he can organize them. Let's see what happens when he tries to keep track of all of them as they come into his throne room.

◆ Since there are so many royal commissioners, the King asks his two Royal Advisors to help keep count. They each have a large notepad. What do you think they are writing on their notepad? (They are writing down numbers as the commissioners come in; they are adding up all of the people, etc.) It says in the book they are making tallies on the notepads. How did the Royal Advisors *create a shortcut* for counting all of the commissioners? (They can make a tally for each person that comes in, then count up the tally marks at the end.)

◆ The King is trying to count the commissioners but his daughter, the Princess, comes in and he loses count when he gives her a hug. When the King asks the First Royal Advisor how many commissioners he counted, he shows the King the notepad. It has tally marks on it for every commissioner who came in the door and then he circled the tally marks in twos. How could he figure out how many commissioners came in? (He could count them by twos.)

- ◆ The King is confused by the First Royal Advisor's method so he asks the Second Royal Advisor to show him his notepad. He also made a tally mark for every commissioner but he circled the tally marks in groups of five. How could he figure out how many commissioners came in? (He could count them by fives.)
- ◆ The King is confused by the Second Royal Advisor's method. The Princess says she has a way to figure out how many commissions there are. She tells the Royal Organizer to put the commissioners in rows so there are 10 in each row. There were 4 rows of 10 and 7 commissioners left over. How could she figure out how many commissioners came in? (She could count them by tens and then add in the ones that are left over.)
- ◆ Let's see how the Princess figures it out. She walks down the rows and says, "10, 20, 30, 40. Plus 7 more makes 47." How did she *create a shortcut* for counting the commissioners? (She put them in groups of 10 so she could count by tens. Then it was easy to add 7 more.)
- ◆ Good problem solvers not only *can create a shortcut* but they know they *can look for repeated calculations.* In order to figure out how many commissioners in all, which is the problem the King is trying to solve, the First Royal Advisor can look at his notepad and look for an opportunity to use a repeated calculation. That means he will do the same thing over and over. Since he circled the tallies by twos, what is his repeated calculation? (He can count by twos and then add in the leftover tally mark.) He counted by twos until he reached 46 and then added 1 more to make 47.
- ◆ Now the Second Royal Advisor can look at his notepad and look for an opportunity to use a repeated calculation. Since he circled the tallies by fives, what is his repeated calculation? (He can count by fives and then add in the two leftover tally marks.) He counted by fives until he reached 45 and then added 2 more to make 47.

You can use this book as a springboard for a discussion about looking for shortcuts in a problem such as putting numbers into groups of twos, fives or tens. Show examples of how you can look for opportunities to repeat calculations in addition word problems or when counting groups of numbers.

Sir Cumference and All the King's Tens by Cindy Neuschwander

In this book, *Sir Cumference and All the King's Tens* (2009), Lady Di of Ameter and her husband, Sir Cumference, plan a surprise party for the King. As more guests arrive, Lady Di tries to count the group so she can plan the meal. She is finally able to keep count when the people are put into groups of ten and then groups of a hundred and finally into groups of a thousand. The groups are put into tents with the amount of people labeled on a flag for the 9,000, the 900, the 90 and the 9 people in each tent. Your students can learn about place value into the thousands and look for repeated calculations as they follow along with the story.

Number and Operations in Base Ten 2.NBT

Understand place value.

Read and write numbers to 1,000 using base ten numerals, number names and expanded form.

As you read the book, engage students in discussion and activities by asking the following questions based on the *I Can* statements:

◆ In this story, Lady Di of Ameter and Sir Cumference decide to plan a surprise party for the King and they invite people from all over Angleland. As the hundreds of guests arrive, Sir Cumference tries to get them out of the castle and into the meadow so the servants can prepare for the party. But Lady Di needs to know the exact number of guests so she can be sure there is enough food for everyone.

◆ Lady Di tries several ways to count all of the people, including lining them all up one by one and putting them in circles. None of these ways prove to be efficient or accurate until Sir Cumference starts to line up the people in rows of ten. We *can create a shortcut* in order to be efficient and accurate when we are solving a problem. Why is this a good shortcut? (They can count the people by tens; it does not take as long to count by tens as it does to count each person one at a time.)

◆ On page 14, we see the groups of ten stretched across the meadow. The shortcut of putting the people into rows of ten could help Sir Cumference be accurate but why is it not so efficient with so many people? (It will take a long time to keep counting by tens.) How could the rows of ten be grouped again in order to *create a shortcut* that would be efficient and accurate? (They could be grouped into hundreds.)

◆ We see on page 16 that Lady Di puts 10 rows of 10 people together to form groups of 100. The counting goes more quickly and she counts 9 groups of 100, 8 rows of 10 and 1 row with 7 people. Let's write that in expanded form to help us find the sum: $900 + 80 + 7 = 987$.

◆ But more people continue to arrive and Lady Di has to join them with the existing groups. She tells the 25 people who show up first to have 3 of their group join the line of 7 to make a 10 and the rest should make rows of 10. Now we have to figure out how many people there are based on the revised groups of hundreds, tens and ones. We have to *pay attention to the details while we think of the goal of the problem.* We know the goal of the problem is to find the total number of people so what operation will we use? (We will use addition.) Now we have to think about the details. Lady Di asked 3 of the 25 people to join the group of 7. Let's decompose 25 into $22 + 3$ and join the 3 to the previous total of 987. Why did she ask 3 people to join the row of 7? (They can make a row of 10 because $7 + 3 = 10$.) When we add 987 and 3, we know we are going to make a new 10, so we have 990 people at this point. There are still 22 people to join the whole group. We *can create a shortcut* so we can use mental math by decomposing 22 into $10 + 12$. How does decomposing 22 this way help us with our addition problem? (We can add 10 to 990 and get 1,000 and then add 12 more and we get 1,012.)

◆ Yes, there are now 1,012 people at the party. It is very hot so Sir Cumference sets up tents of different sizes to provide shade for the guests. The tents hold a certain number of people so flags are posted on the tents to indicate the maximum number of people. The smallest tent can hold only 9 people, the tent to its right can hold 90, the tent to its right can hold 900 and the tent to its right can hold up to 9,000 people.

◆ The entire city of Camelot just arrived and Sir Cumference organizes them into groups out in the meadow so Lady Di can count them and then add the number to 1,012. Once again the counting goes quickly and she counts 8 groups of 1,000, 9 groups of 100, 8 rows of 10 and 1 row with 7 people. Let's write that in expanded form to help us find the sum: $8,000 + 900 + 80 + 7 = 8,987$.

◆ Now we have to figure out how many people there are based on the new group of thousands, hundreds, tens and ones. We have to *pay attention to the details while we think of the goal of the problem.* We know the goal of the problem is to find the total number of people so what operation will we use? (We will use addition.) Now we have to think about the details. Lady Di asks all of the guests to get into the tents, starting with the smallest tent, so they can surprise the King when he arrives. The number of people in the first group is 1,012 and the number of people in the second group is 8,987. How many fit into the first tent? (Nine can fit in the first tent). If we start with the ones in each group, the people who do not make a complete row, how many are left from each group? (If we put the 2 people from the 1,012 group and 7 people from the 8,987 group into the tent there would be 1,010 left in the first group and 8,980 in the second group.)

◆ We *can create a shortcut* so we can use mental math by keeping track of each place value. Let's put the numbers back into expanded form so we can see how many people would go into each tent from the two groups of people. The first number would now be $1,000 + 10$ and the second number would be $8,000 + 900 + 80$. The next largest tent holds 90 people. We do not have any more ones so we have to go to the tens in each group. If they all go into the tent, how many are left from each group? (If we put the 10 people from the first group and the 80 people from the second group into the tent there would be 1,000 people left in the first group and 8,900 left in the second group.)

◆ The next tent holds 900 people so which group goes into that tent? (The 900 people from the second group fit in that tent.) The last tent holds 9,000 people so which group goes into that tent? (The 1,000 people from the first group and the 8,000 people from the second group fit in that tent.) Now we can get back to the goal of our problem: How many people are at the party and how do you know? (There are 9,999 people at the party; they all fit into the tents and that is how many people fit into the tents altogether.)

Your students can create additional stories about Lady Di and Sir Cumference using large numbers and opportunities to read, write and group numbers in the thousands using properties of place value. They can also look for repeated calculations and shortcuts as they group the numbers.

What Does This Standard Mean for Grades 3–5 Problem Solvers?

Students who are successful at SMP 8 are familiar with efficient strategies and shortcuts that can be applied to various word problems. They recognize the types of problem solving situations for the four operations (see Tables A1.1 and A1.2 in Appendix) based on the structure of the problem and if calculations are repeated. They look for opportunities to apply formulas and can make conclusions about the calculations used in the problem solving process and the solution.

In order to fully apply SMP 8 when approaching a word problem, students in intermediate grades should be able to take ownership of their procedures by using the following *I Can* statements:

◆ *I can look for repeated calculations, general methods or shortcuts while solving the problem.*
◆ *I can recognize when a general formula exists to help solve the problem.*
◆ *I can make conclusions about my results as I work through a possible solution.*

Math Appeal by Greg Tang

Math Appeal (2003) by Greg Tang is one of his many books that poses riddles for the reader to solve by using patterns within the language of the poems and within the pictures on every page. Students should look for number patterns, such as grouping numbers and making tens. Other clues direct students to add rows in an array as if all objects are present then subtract the missing objects. Your students can also find patterns using arrays and equal groups to practice their multiplication facts.

Operations and Algebraic Thinking 3.OA

Represent and solve problems involving multiplication and division.
Interpret products of whole numbers, e.g., interpret 5×7 as the total number of objects in 5 groups of 7 objects each.

As you read the poems throughout the book, engage students in discussion and activities by asking the following questions based on the *I Can* statements:

◆ The author of this book, Greg Tang, wants children to become clever, creative thinkers so he uses language and the illustrations in his books to promote problem solving strategies. We know we *can look for repeated calculations, general methods or shortcuts while solving a problem* so let's see how we can practice these skills as we read his book.
◆ The first poem, Square Deal, is about a kite with a pattern of squares. Greg Tang likes to include clues in his poems that tell the reader how to efficiently count objects. This poem tells us,

"How many squares? Let me see, it's best to add diagonally!" If we look at the pattern on the kite, why is it more efficient to count diagonally? (The squares are in groups of five if you look at the diagonal pattern; it is quicker to count by fives.)

◆ So if we count by fives, how many squares are on the kite? (There are 25 squares.) It was much more efficient to count by fives. If we wanted to create a multiplication equation for that problem, what would it be? ($5 \times 5 = 25$.) This is an example of a perfect square. When we multiply a number by itself the result is a perfect square.

◆ The next poem is Boston Pea Party. The clue in this poem is at the end like the first poem, "Can you count up all the peas? With 11s it's a breeze!" If we look at the illustration of the sets of peas, where do we find the 11s indicated in the poem? (Some peapods have 6 peas and some have 5 peas; we know $5 + 6 = 11$.) Now we know the sets of peapods make 11, so now what do we have to do if we want to count all of the peas on the page? (We count by 11s; we can multiply to find the total.) If we wanted to create a multiplication equation for that problem, what would it be? ($4 \times 11 = 44$.) What do the numbers represent in that equation? (There are 4 groups of 11 peas.)

◆ So far we have been using *repeated calculations* as a *shortcut* to count the objects in the illustrations. We were able to group objects to use the method of multiplying equal groups. Now we will read another poem and, although we will still look for *repeated calculations* as a *shortcut*, there will be a different type of *general method*.

◆ The poem is called Frog-Gone! The clue at the end of the poem states, "Here's a tip to help you add, don't ignore a lily pad!" That seems odd, to count the frogs that are not even there. But let's try it and see why this clue is important. How many frogs would there be if we counted every lily pad? (There would be 25.) Okay, there are 5 rows of lily pads and 5 lily pads in each row. That is just like the pattern in the first poem. We know $5 \times 5 = 25$. But there are 6 frogs missing. How would we find out the number of frogs that are on the lily pads? (We would subtract; we would use the equation $25 - 6 = 19$.) So it is more efficient to do the quick multiplication then simply subtract to get 19 rather than count all of the frogs by ones. There will be other poems that will use the same method so this will become another *general method* in our problem solving bag of tricks.

◆ In this next poem, Red Hot Chili Peppers, some of the chili peppers are arranged in a particular shape. The clue in the poem states, "Group the chilies in the square, Add the rest and you'll be there!" If we look at the illustration and imagine the chili peppers in the middle forming a square, how many are there and how do you know? (There are 16; there are 4 rows of 4.) If we wanted to create a multiplication equation for that problem, what would it be? ($4 \times 4 = 16$.) This is also an example of a perfect square. When we multiply a number by itself the result is a perfect square. There are a few chili peppers that were not part of the square and the clue tells us to "add the rest" and we will have our answer. What is the total number of chili peppers and how do you know? (There are 20; there are 4 outside of the square; $16 + 4 = 20$.)

◆ This poem is called Lucky Clovers. The poem lets us know there are some clovers with 3 leaves and some with 4 leaves and the clue tells us, "Instead of counting row by row, why not add the leaves below!" We are going to use a *shortcut* of *repeated calculations* again. Why is it more efficient to add the leaves in the column and not the row? (There are 10 leaves in each column; it is easier to count by tens.) If we wanted to create a multiplication equation for that problem, what would it be? ($4 \times 10 = 40$.) What do the numbers represent in that equation? (There are 4 groups of 10 leaves.) We can think of this as a *general formula* for these types of problems. There are always equal groups. The first factor represents the total number of groups and the second factor represents the number of objects in each group.

◆ Here is another poem, Rock Stars, with the clue, "Instead of counting one by one, just subtract and you'll be done!" In which other problem did we subtract to help us find the total number of objects? (The poem with the frogs on the lily pads.) Let's review what we did with the frogs on the lily pads. We used a *repeated calculation* as a *shortcut* to count by fives or think about multiplying 5×5. Then we subtracted the number of missing frogs to get our total.

◆ In this illustration, how many starfish are in each complete row? (There are 6 in each row.) How many rows are there? (There are 6 rows.) What is our repeated calculation if every row had 6 starfish? (We would add 6 six times; we would multiply 6×6.) So how many starfish would there be? (There would be 36.) If we wanted to create a multiplication equation for that problem, what would it be? ($6 \times 6 = 36$.) This is another example of a perfect square. When we multiply a number by itself the result is a perfect square. But we know some of the starfish are missing and the *general formula* for this type of problem is to add or multiply as if all of the objects are there and then subtract the ones that are missing. How would we find out the number of starfish using this formula? (We would subtract; we would use the equation $36-6=30$.) So it is more efficient to do the quick multiplication then simply subtract to get 30 rather than count all of the starfish by ones.

There are 16 poems in this book which allow for students to use repeated calculations as shortcuts. Encourage students to find the most efficient way to count the objects before you read the clue at the end of the poem. They should think about the clues in the poems you have already done as models for counting diagonally, vertically and in groups rather than always counting one by one and from left to right. You can also require your students to explain their shortcut to a peer and then be able to write out their explanation.

The Warlord's Beads by Virginia Walton Pilegard

In this book, *The Warlord's Beads* (2001), a boy named Chuan and his father live in the palace of a powerful Chinese warlord. Chuan's father has the job of sorting and counting all of the warlord's treasures, which increase each day. Chuan attempts to help his father count the objects in groups of ten but they realize they cannot keep track of all of the piles of ten with their fingers and toes. Chuan sees some thin branches on which he strings ten beads so he can move the beads back and forth to keep track of the large numbers.

Number and Operations in Base Ten 4.NBT

Generalize place value understanding for multi-digit whole numbers.
Recognize that in a multi-digit whole number, a digit in one place represents ten times what is represents in the place to its right.

As you read the book, engage students in discussion and activities by asking the following questions based on the *I Can* statements:

◆ This book is about a boy named Chuan who lived with his father in the warlord's palace in China long ago. Chuan's father works for the warlord with the duty of sorting and counting all of the warlord's many treasures. We can see on the fifth page many boxes and statues in the counting room and Chuan's father sitting with his eyes closed. Chuan finds out his father has been accused by the warlord of stealing his treasures because the number of treasures his father counts each day goes up and down. They realize the problem is that his father loses count when someone interrupts.

◆ On the tenth page, we see Chuan and his father trying out some ideas. What does it look like they're doing? (They are stacking boxes; they are making groups of ten boxes; Chuan is using his fingers to keep track of the groups of ten.) We know to look *for repeated calculations, general methods or shortcuts while solving a problem.* It is a *shortcut* to make stacks of ten boxes and count them by tens. But we see on the next page they have quickly reached ten stacks of ten boxes and Chuan says he is out of fingers to keep track. His father tells him to use his toes! But he does seem to be coming up with a method. It says on this page, "One hundred ten—one toe and one finger." What do his toe and finger represent? (He used his fingers to keep track of the tens, then when he got to one hundred he lifted one toe for the one hundred; he is starting over counting tens so one finger represents one ten.)

◆ Good problem solvers *can make conclusions about their results as they work through a possible solution.* We can see Chuan has a good idea but what conclusions can you make at this point as he is using this method? (He will run out of toes also; he will have a hard time lifting each toe and might get mixed up or lose count.) We can see on page 19 that Chuan has lost track of how many fingers and toes he held up when the warlord's three sons came into the counting room waving their switches, which are thin branches, and started chasing Chuan.

◆ But on the next page we see Chuan pick up one of the switches and push some beads onto it. What does it look like he is doing? (Chuan is using the beads to keep track of the groups of ten instead of his fingers and toes.) We know to look *for repeated calculations, general methods or shortcuts while solving a problem.* On the next page we see he has threaded ten beads onto each of the three switches. He is still using his original *shortcut* of stacking boxes and counting them by tens but he will use the beads to keep track now.

◆ Let's look at the beads that have been moved to the top of each of the switches so far. (See Figure 8.1.)

◆ He started with the green beads on the switch on the right. We see one bead pushed to the top so how many single boxes, or ones,

Figure 8.1

does he have? (He has one box.) We see three beads pushed to the top of the switch to its left. How many stack of boxes, or tens, does he have? (He has three stacks of ten boxes.) What do you think the next switch represents? Think about what happens after he moves all ten beads on the tens switch. (That switch represents hundreds.) We *can make conclusions about our results as we work through a possible solution.* Let's look at Chuan's results so far. How many treasures has he counted and how do you know? (He has counted 131; he has one bead moved on the ones, three beads moved on the tens and one bead moved on the hundreds.)

◆ How does this method use a *shortcut* of *repeated calculations*? (He can keep moving the beads up on the first switch until all ten beads are at the top, then he moves up one bead on the tens switch to show he counted ten boxes. Then he goes back to the first switch and repeats the same process until he gets another ten. When he has ten tens he moves one bead on the hundreds switch and starts over again with the first switch.)

◆ By the end of the day, Chuan and his father successfully counted all of the warlord's treasures using the method with the beads on the switches. His method of keeping track of the ones, tens and hundreds worked! You can see on the last page of the book that Chuan used a counting frame, which came before the *abacus*, which was a type of "calculator" used by the fourteenth century in Asia.

Your students can create their own counting beads using pipe cleaners and cereal or plastic beads. Challenge them to go beyond the hundreds place by using additional sets of pipe cleaners and beads so they can explain the relationship between the value of each digit and the number to its right.

Anno's Magic Seeds by Mitsumasa Anno

The book *Anno's Magic Seeds* (1995) tells the story of Jack, who meets a wizard who gives him two magic seeds. If he eats one seed, he will not be hungry for a whole year. The wizard tells him to plant the other seed and when it blooms it will provide two fruits with a seed in each and he can start the process all over again. After several years of doing this, Jack buries both seeds in the ground and eats something else throughout the winter. The next spring he ends up with two plants and four fruits. He eats one and plants the remaining three, starting a growing pattern of seeds produced each year.

Operations and Algebraic Thinking 5.OA

Analyze patterns and relationships.

Generate two numerical patterns using two given rules. Identify apparent relationships between corresponding terms. Form ordered pairs consisting of corresponding terms from the two patterns, and graph the ordered pairs on a coordinate plane.

As you read the book, engage students in discussion and activities by asking the following questions based on the *I Can* statements:

◆ You may know the story of *Jack and the Beanstalk* and how the beanstalk grew from the magic seeds Jack buried in the ground. This story is about Jack and another type of magic seed. A wizard gives him two seeds, one to eat and one to bury in the ground. If he eats the one seed he will not be hungry for a whole year. He follows the wizard's instructions and eats the first seed and buries the second seed. He continues doing this for several years until one day he gets an idea.

◆ We see on page 10 that he is burying both seeds. What do you think will happen as a result of doing this? (He will have to find something else to eat because he buried both seeds; he will have two plants that grow in the spring.) Yes, we see on the next page that he did have two plants in the spring and there are four fruits. On page 12 he buried three seeds and he had three plants in the spring with six fruits. We *can look for repeated calculations, general methods or shortcuts while solving a problem.* Let's try to figure out how many fruits he will have each spring based on the number of seeds he plants. We can make a table with the information we have at this point in the story and use it to see if we notice any *repeated calculations* we can use as a *shortcut* to figure out how many fruits he will have in future years. (See Table 8.1.)

◆ Based on the numbers in the table, what do you notice about the numbers in the first column? (They are counting by ones; each number is one more than the number before it.) What do you notice about the numbers in the second column? (They are counting by twos; each number is two more than the number before it.) Do you notice a relationship between the numbers in each row? (The first number is smaller than the second number; the second number is twice as much as the first number.)

Table 8.1

Number of Seeds Planted	Number of Fruits
1	2
2	4
3	6
4	?

- ◆ We can also *recognize when a general formula exists to help solve a problem.* Let's see if we can make a formula with this information. First we can label the first column as x and the second column as y. Now think about the idea one of you had about the relationship between the numbers in the first column and the second column: the second number is twice as much as the first number. How could we translate that into a formula using x and y? (We could take twice x and it would equal y; $2x = y$.) How can we apply that formula to fill in the missing information in the table? (We can put 4 into the formula to replace x and we get $2 \times 4 = 8$).
- ◆ On page 13, we see that he had ten seeds from the ten fruits. How many seeds did he bury to get ten fruits? We *can make conclusions about our results as we work through a possible solution.* First we have to identify what we are looking for in the formula. Are we looking for the x or the y in the formula and how do you know? (We are looking for the x because we know the number of fruits, but we do not know the number of seeds he planted.) What is our new equation using the formula? ($2x = 10$.) We can use the formula to solve the equation as well as the pattern in the table. How many seeds did he plant? (He planted five seeds.)

Continue filling in the table as you read the story. Your students can use the pattern and the formula to figure out the number of fruits and how many seeds he buried. When your students are ready, use the numbers in the table as ordered pairs on a coordinate plane using only Quadrant 1 since there are only positive values.

Wrapping It Up

Find additional examples of word problems that allow students to create shortcuts and look for repeated calculations to help them arrive at their solution. Use a think-aloud method to model how to read word problems for details that can help with the solution process. Once your students are able to do this with one-step word problems, model how to look for repeated calculations, shortcuts and details to help solve two-step word problems.

Next Steps

Once your students have been exposed to the Standards for Mathematical Practice using some of the children's literature from this book, choose other books that can lend themselves to the SMP as well as to the content standards. Share ideas with your colleagues or start a list with other teachers in an online environment. Continue improving the culture of mathematical problem solving in your school by introducing grade-level or school-wide problem solving situations, such as figuring out the area and perimeter of the playground, graphing the amount of garbage generated by the school each day or establishing a school-wide fundraiser conducted by the students.

Include a poster in your classroom with the eight SMP and the *I Can* statements so students can refer to them as they progress through problem solving activities throughout the year. Students should be able to explain which SMP they are using and why. Share explanations of the problem solving process with parents in newsletters, on the school or classroom website, at Open House or Parent–Teacher Conferences and even in the math work that students bring home. Create extensions of the classroom activities to be sent home so students can work with their family to solve problems using the SMP.

Encourage your school to begin housing resources for incorporating problem solving and the Common Core State Standards for Mathematics, such as professional books, periodicals, DVDs, games, center activities and children's literature that can be used with math content areas. Collaborate with other teachers to find ways to integrate other content areas into math, such as art, music and physical education. Students can study concepts in geometry while working on an art project, explore patterns and counting while they plan an instrument in music and measure distance and time in physical education class.

I hope this book has helped illuminate the eight SMP as well as some of the content standards through the use of children's literature. Perhaps this book served as an introduction to the language used in the Common Core State Standards for Mathematics document or provided ideas as to how the SMP can be broken down for younger students. Please persist in teaching your students the importance of the process of problem solving rather than simply focusing on which student can provide the correct answer.

Appendix -3rd-5th grade

Table A1.1

	Result Unknown	Change Unknown	Start Unknown
Add to	Two bunnies sat on the grass. Three more bunnies hopped there. How many bunnies are on the grass now? $2+3=n$	Two bunnies were sitting on the grass. Some more bunnies hopped there. Then there were 5 bunnies. How many bunnies hopped over to the first 2? $2+n=5$	Some bunnies were sitting on the grass. Three more bunnies hopped there. Then there were 5 bunnies. How many bunnies were on the grass before? $n+3=5$
Take from	Five apples were on the table. I ate 2 apples. How many apples are on the table now? $5-2=n$	Five apples were on the table. I ate some apples. Then there were 3 apples. How many apples did I eat? $5-n=3$	Some apples were on the table. I ate 2 apples. Then there were 3 apples. How many apples were on the table before? $n-2=3$
	Total Unknown	**Addend Unknown**	**Both Addends Unknown**
Put Together/ Take Apart	Three red apples and 2 green apples are on the table. How many apples are on the table? $3+2=n$	Five apples are on the table. Three are red and the rest are green. How many apples are green? $3+n=5, 5-3=n$	Grandma has 5 flowers. How many can she put in her red vase and how many in her blue vase? $5=0+5, 5=5+0$ $5=1+4, 5=4+1$ $5=2+3, 5=3+2$
	Difference Unknown	**Bigger Unknown**	**Smaller Unknown**
Compare	("How many more?" version): Lucy has 2 apples. Julie has 5 apples. How many more apples does Julie have than Lucy? ("How many fewer?" version): Lucy has 2 apples. Julie has 5 apples. How many fewer apples does Lucy have than Julie? $2+n=5, 5-2=n$	(Version with "more"): Julie has 3 more apples than Lucy. Lucy has 2 apples. How many apples does Julie have? (Version with "fewer"): Lucy has 3 fewer apples than Julie. Lucy has 2 apples. How many apples does Julie have? $2+3=n, 3+2=n$	(Version with "more"): Julie has 3 more apples than Lucy. Julie has 5 apples. How many apples does Lucy have? (Version with "fewer"): Lucy has 3 fewer apples than Julie. Julie has 5 apples. How many apples does Lucy have? $5-3=n, n+3=5$

Table A1.2

	Unknown Product	Group Size Unknown ("How Many in Each Group?" Division)	Number of Groups Unknown ("How Many Groups?" Division)
	$3 \times 6 = ?$	$3 \times ? = 18$, and $18 \div 3 = ?$	$? \times 6 = 18$, and $18 \div 6 = ?$
Equal Groups	There are 3 bags with 6 plums in each bag. How many plums are there in all? *Measurement example.* You need 3 lengths of string, each 6 inches long. How much string will you need altogether?	If 18 plums are shared equally into 3 bags, then how many plums will be in each bag? *Measurement example.* You have 18 inches of string, which you will cut into 3 equal pieces. How long will each piece of string be?	If 18 plums are to be packed 6 to a bag, then how many bags are needed? *Measurement example.* You have 18 inches of string, which you will cut into pieces that are 6 inches long. How many pieces of string will you have?
Array, Area	There are 3 rows of apples with 6 apples in each row. How many apples are there? *Area example.* What is the area of a 3 cm by 6 cm rectangle?	If 18 apples are arranged into 3 equal rows, how many apples will be in each row? *Area example.* A rectangle has area 18 square centimeters. If one side is 3 cm long, how long is a side next to it?	If 18 apples are arranged into equal rows of 6 apples, how many rows will there be? *Area example.* A rectangle has area 18 square centimeters. If one side is 6 cm long, how long is a side next to it?
Compare	A blue hat costs $6. A red hat costs 3 times as much as the blue hat. How much does the red hat cost? *Measurement example.* A rubber band is 6 cm long. How long will the rubber band be when it is stretched to be 3 times as long?	A red hat costs $18 and that is 3 times as much as a blue hat costs. How much does a blue hat cost? *Measurement example.* A rubber band is stretched to be 18 cm long and that is 3 times as long as it was at first. How long was the rubber band at first?	A red hat costs $18 and a blue hat costs $6. How many times as much does the red hat cost as the blue hat? *Measurement example.* A rubber band was 6 cm long at first. Now it is stretched to be 18 cm long. How many times as long is the rubber band now as it was at first?
General	$a \times b = ?$	$a \times ? = p$, and $p \div a = ?$	$? \times b = p$, and $p \div b = ?$

References

Aboff, Marcie. 2010. *If You Were a Triangle*. Minneapolis, MN: Picture Window Books.

Adler, David A. 2010. *Fractions, Decimals and Percents*. New York: Holiday House.

Adler, David A. 2012. *Perimeter, Area and Volume*. New York: Holiday House.

Anno, Mitsumasa. 1995. *Anno's Magic Seeds*. New York: Penguin Putnam.

Burns, Marilyn. 1994. *The Greedy Triangle*. New York: Scholastic, Inc.

Burns, Marilyn. 1997. *Spaghetti and Meatballs for All!* New York: Scholastic, Inc.

Calvert, Pam. 2006. *Multiplying Menace: The Revenge of Rumpelstiltskin*. Watertown, MA: Charlesbridge.

Carle, Eric. 1972. *Rooster's Off to See the World*. New York: Simon & Schuster.

Common Core State Standards Initiative (CCSSI). 2010. *Common Core State Standards*. Washington, DC: National Governors Association Center for Best Practices and Council of Chief State School Officers.

Dahl, Michael. 2006. *Bunches of Buttons: Counting by Tens*. Minneapolis, MN: Picture Window Books.

Dodds, Dayle Ann. 2007. *Full House: An Invitation to Fractions*. Cambridge, MA: Candlewick Press.

Friedman, Aileen. 1994. *The King's Commissioners*. New York: Scholastic, Inc.

Garland, Michael. 2007. *How Many Mice?* New York: Dutton Children's Books.

Giganti, Paul. 1992. *Each Orange Had 8 Slices*. New York: Greenwillow Books.

Hutchins, Pat. 1986. *The Doorbell Rang*. New York: Greenwillow Books.

Jenkins, Steve. 2004. *Actual Size*. Boston, MA: Houghton Mifflin.

Jonas, Ann. 1995. *Splash!* New York: Greenwillow Books.

Leedy, Loreen. 1997. *Measuring Penny*. New York: Henry Holt and Co.

McNamara, Margaret. 2007. *How Many Seeds in a Pumpkin?* New York: Schwartz & Wade Books.

Murphy, Stuart J. 1998. *Lemonade for Sale*. New York: Harper Collins Publishers.

Murphy, Stuart J. 2002. *Bigger, Better, Best!* New York: Harper Collins Publishers.

Murphy, Stuart J. 2002. *Racing Around.* New York: Harper Collins Publishers.

Murphy, Stuart J. 2004. *Earth Day—Hooray!* New York: Harper Collins Publishers.

Murphy, Stuart J. 2005. *Polly's Pen Pal.* New York: Harper Collins Publishers.

Murphy, Stuart J. 2006. *Mall Mania.* New York: Harper Collins Publishers.

Myller, Rolf. 1962. *How Big is a Foot?* New York: Dell Publishing.

Nagda, Ann Whitehead. 2007. *Cheetah Math: Learning about Division from Baby Cheetahs.* New York: Henry Holt and Co.

Nagda, Ann Whitehead, and Cindy Bickel. 2000. *Tiger Math: Learning to Graph from a Baby Tiger.* New York: Scholastic, Inc.

Nagda, Ann Whitehead, and Cindy Bickel. 2002. *Chimp Math: Learning about Time from a Baby Chimpanzee.* New York: Henry Holt and Co.

Neuschwander, Cindy. 2001. *Sir Cumference and the Great Knight of Angleland.* Watertown, MA: Charlesbridge.

Neuschwander, Cindy. 2003. *Sir Cumference and the Sword in the Cone.* New York: Scholastic, Inc.

Neuschwander, Cindy. 2007. *Patterns in Peru: An Adventure in Patterning.* New York: Henry Holt and Co.

Neuschwander, Cindy. 2009. *Sir Cumference and All the King's Tens.* Watertown, MA: Charlesbridge.

Neuschwander, Cindy. 2012. *Sir Cumference and the Viking's Map.* Watertown, MA: Charlesbridge.

Pallotta, Jerry. 1999. *The Hershey's Milk Chocolate Fractions Book.* New York: Scholastic, Inc.

Pallotta, Jerry. 2002. *Hershey's Milk Chocolate Weights and Measures.* New York: Scholastic, Inc.

Pallotta, Jerry. 2003. *Count to a Million.* New York: Scholastic, Inc.

Pallotta, Jerry. 2004. *Hershey's Kisses from Addition to Multiplication.* New York: Scholastic, Inc.

Pilegard, Virginia. 2000. *The Warlord's Puzzle.* Gretna, LA: Pelican Publishing Co.

Pilegard, Virginia. 2001. *The Warlord's Beads.* Gretna, LA: Pelican Publishing Co.

Pilegard, Virginia. 2004. *The Warlord's Kites.* Gretna, LA: Pelican Publishing Co.

Pinczes, Elinor. 1995. *A Remainder of One.* New York: Houghton Mifflin Co.

Pinczes, Elinor. 2001. *Inchworm and a Half.* New York: Houghton Mifflin Co.

Reid, Margarette. 1990. *The Button Box.* New York: Penguin Books.

Schwartz, David M. 1999. *If You Hopped Like a Frog...* New York: Scholastic, Inc.

Scieszka, Jon. 1995. *The Math Curse.* New York: Penguin Books.

Sturges, Philemon. 1995. *Ten Flashing Fireflies.* New York: North-South Books.

Tang, Greg. 2003. *Math Appeal.* New York: Scholastic, Inc.

Tang, Greg. 2003. *MATH-terpieces.* New York: Scholastic, Inc.

Viorst, Judith. 1978. *Alexander, Who Used to be Rich Last Sunday.* New York: Scholastic, Inc.

Young, Ed. 1992. *Seven Blind Mice.* New York: Scholastic, Inc.